Twisted Tensor Products Related to the Cohomology of the Classifying Spaces of Loop Groups

Memoirs
of the
American Mathematical Society

Number 849

Twisted Tensor Products Related to the Cohomology of the Classifying Spaces of Loop Groups

Katsuhiko Kuribayashi
Mamoru Mimura
Tetsu Nishimoto

March 2006 • Volume 180 • Number 849 (third of 5 numbers) • ISSN 0065-9266

American Mathematical Society
Providence, Rhode Island

2000 *Mathematics Subject Classification.* Primary 55T20, 57T30, 55S05.

Library of Congress Cataloging-in-Publication Data

Kuribayashi, Katsuhiko, 1963–
  Twisted tensor products related to the cohomology of the classifying spaces of loop groups / Katsuhiko Kuribayashi, Mamoru Mimura, Tetsu Nishimoto.
    p. cm. — (Memoirs of the American Mathematical Society, ISSN 0065-9266 ; no. 849)
  Includes bibliographical references.
  ISBN 0-8218-3856-3 (alk. paper)
    1. Spectral sequences (Mathematics) 2. Topology. 3. Cohomology operations. I. Mimura, M. (Mamoru), 1938– II. Nishimoto, Tetsu, 1969– III. Title. IV. Series.

QA3.A57 no. 849
[QA612.8]
510 s—dc22
[514′.2]   2005057159

Memoirs of the American Mathematical Society

This journal is devoted entirely to research in pure and applied mathematics.

Subscription information. The 2006 subscription begins with volume 179 and consists of six mailings, each containing one or more numbers. Subscription prices for 2006 are US$624 list, US$499 institutional member. A late charge of 10% of the subscription price will be imposed on orders received from nonmembers after January 1 of the subscription year. Subscribers outside the United States and India must pay a postage surcharge of US$31; subscribers in India must pay a postage surcharge of US$43. Expedited delivery to destinations in North America US$35; elsewhere US$130. Each number may be ordered separately; *please specify number* when ordering an individual number. For prices and titles of recently released numbers, see the New Publications sections of the *Notices of the American Mathematical Society*.

Back number information. For back issues see the *AMS Catalog of Publications*.

Subscriptions and orders should be addressed to the American Mathematical Society, P. O. Box 845904, Boston, MA 02284-5904, USA. *All orders must be accompanied by payment.* Other correspondence should be addressed to 201 Charles Street, Providence, RI 02904-2294, USA.

Copying and reprinting. Individual readers of this publication, and nonprofit libraries acting for them, are permitted to make fair use of the material, such as to copy a chapter for use in teaching or research. Permission is granted to quote brief passages from this publication in reviews, provided the customary acknowledgment of the source is given.

Republication, systematic copying, or multiple reproduction of any material in this publication is permitted only under license from the American Mathematical Society. Requests for such permission should be addressed to the Acquisitions Department, American Mathematical Society, 201 Charles Street, Providence, Rhode Island 02904-2294, USA. Requests can also be made by e-mail to `reprint-permission@ams.org`.

Memoirs of the American Mathematical Society is published bimonthly (each volume consisting usually of more than one number) by the American Mathematical Society at 201 Charles Street, Providence, RI 02904-2294, USA. Periodicals postage paid at Providence, RI. Postmaster: Send address changes to Memoirs, American Mathematical Society, 201 Charles Street, Providence, RI 02904-2294, USA.

© 2006 by the American Mathematical Society. All rights reserved.
Copyright of this publication reverts to the public domain 28 years
after publication. Contact the AMS for copyright status.
This publication is indexed in *Science Citation Index*®, *SciSearch*®, *Research Alert*®, *CompuMath Citation Index*®, *Current Contents*®/*Physical, Chemical & Earth Sciences*.
Printed in the United States of America.

∞ The paper used in this book is acid-free and falls within the guidelines
established to ensure permanence and durability.
Visit the AMS home page at `http://www.ams.org/`

10 9 8 7 6 5 4 3 2 1   11 10 09 08 07 06

Contents

1. Introduction — 1
2. The mod 2 cohomology of $BLSO(n)$ — 9
3. The mod 2 cohomology of BLG for $G = Spin(n)$ ($7 \leq n \leq 9$) — 10
4. The mod 2 cohomology of BLG for $G = G_2, F_4$ — 13
5. A multiplication on a twisted tensor product — 15
6. The twisted tensor product associated with $H^*(Spin(N); \mathbb{Z}/2)$ — 35
7. A manner for calculating the homology of a DGA — 39
8. The Hochschild spectral sequence — 45
9. Proof of Theorem 1.6 — 50
10. Computation of a cotorsion product of $H^*(Spin(10); \mathbb{Z}/2)$ and the Hochschild homology of $H^*(BSpin(10); \mathbb{Z}/2)$ — 63
11. Proof of Theorem 1.7 — 70
12. Proofs of Proposition 1.9 and Theorem 1.10 — 73
13. Appendix — 77

Bibliography — 83

Abstract

Let G be a compact, simply connected, simple Lie group. By applying the notion of a twisted tensor product in the senses of Brown as well as of Hess, we construct an economical injective resolution to compute, as an algebra, the cotorsion product which is the E_2-term of the cobar type Eilenberg-Moore spectral sequence converging to the cohomology of classifying space of the loop group LG. As an application, the cohomology $H^*(BLSpin(10); \mathbb{Z}/2)$ is explicitly determined as an $H^*(BSpin(10); \mathbb{Z}/2)$-module by using effectively the cobar type spectral sequence and the Hochschild spectral sequence, and further, by analyzing the TV-model for $BSpin(10)$.

2000 *Mathematics Subject Classification.* 55T20, 57T30, 55S05.

Key words and phrases. loop groups, twisted tensor products, the Hochschild spectral sequence, the bar and cobar type Eilenberg-Moore spectral sequences, TV-models.

The first author was partially supported by the grant-in-aid for Scientific Research (C)14540095 from JSPS. The second author was partially supported by the grant-in-aid of JSPS, ♯ 1240025.

Received by the editor September 18, 2003.

1. Introduction

In the mid-1970s, Kono, Mimura and Shimada ([**23**],[**24**],[**22**]) have determined the mod p cohomology groups of $BPU(3)$ and BF_4 for $p = 3$, and BE_6 and BE_7 for $p = 2$, using the Rothenberg-Steenrod spectral sequence

$$\mathrm{Cotor}_{H^*(G;\mathbb{Z}/p)}(\mathbb{Z}/p, \mathbb{Z}/p) \Longrightarrow H^*(BG; \mathbb{Z}/p).$$

The notion of a twisted tensor product in the sense of Brown [**3**] plays an important role in their consideration. Roughly speaking, the twisted tensor product associated with the mod p cohomology A of a Lie group is the tensor product of A and a complex $(\overline{X}_A, d)$ equipped with a perturbed differential, and in addition it is an *economical* injective resolution of $\mathbb{Z}/p$ as a A-comodule. The advantage of the complex $(\overline{X}_A, d)$ is that it is a *manageable* differential graded algebra (DGA for short) which gives the cotorsion product $\mathrm{Cotor}_A(\mathbb{Z}/p, \mathbb{Z}/p)$, namely, the E_2-term of the spectral sequence. The computation of the mod p cohomology of classifying spaces of other Lie groups due to Mimura and Sambe ([**32**],[**33**],[**34**]) has also told us that a twisted tensor product as an injective resolution is relevant to the study of the cohomology via the spectral sequence.

Let G be a compact, connected simple Lie group and LG denote the loop group which is an infinite dimensional manifold consisting of all C^∞-maps from the circle to G. Our interest here lies in computing the mod p cohomology of the classifying space BLG of the loop group LG. In order to compute those cohomologies, we give a DGA structure to the twisted tensor products by also perturbing the algebra structure of tensor product $(A, 0) \otimes (\overline{X}_A, d)$. More precisely, we have the following theorem.

THEOREM 1.1. *Each twisted tensor product, which is constructed by Kono, Mimura, Sambe and Shimada, associated with mod p cohomology A of a Lie group is that of the differential graded algebras $(A, 0)$ and $(\overline{X}_A, d)$ in the sense of Hess* [**15**].

One may perform the computation of the cohomology $H^*(BLG; \mathbb{Z}/p)$ with the aid of the Rothenberg-Steenrod spectral sequence, whose E_2-term is the cotorsion product $\mathrm{Cotor}_{H^*(LG;\mathbb{Z}/p)}(\mathbb{Z}/p, \mathbb{Z}/p)$. As the first step in computing the spectral sequence, we need to construct an economical injective resolution of $\mathbb{Z}/p$ as an $H^*(LG; \mathbb{Z}/p)$-comodule such as the twisted tensor product mentioned above. However, it seems to be hard to carry it out due to infinite number of algebra generators of $H^*(LG; \mathbb{Z}/p)$ and more complicated coalgebra structure $H^*(LG; \mathbb{Z}/p)$ than that of $H^*(G; \mathbb{Z}/p)$. (For the algebra structure of the homology $H_*(LG; \mathbb{Z}/p)$, see [**11**],

[**12**], [**13**] and [**14**].) We then employ other spectral sequences in considering the cohomology $H^*(BLG;\mathbb{Z}/p)$. It is known that there is a homotopy equivalence between BLG and the free loop space $(BG)^{S^1}$ which is the space of all continuous maps from S^1 to BG. Moreover, BLG has the homotopy type of $G \times_{\mathrm{ad}} EG$ which is the total space of the associated bundle

$$G \to G \times_{\mathrm{ad}} EG \to BG$$

to the universal principal G-bundle (see [**5**, Corollary 3.4]). Here the right action of G to itself is the adjoint action

$$\mathrm{ad} : G \times G \to G$$

defined by $\mathrm{ad}(g,h) = h^{-1}gh$. Observe that the projection of the bundle defines an $H^*(BG;\mathbb{Z}/p)$-algebra structure on $H^*(BLG;\mathbb{Z}/p)$, which is the same as that induced by the evaluation map $LG \to G$ at zero (see [**28**, (2.1)]).

We thus see that the following three spectral sequences are also applicable in the study of the cohomology $H^*(BLG;\mathbb{Z}/p)$:

HSS: The Hochschild spectral sequence $\{{}_{HH}E_r^{*,*}, d_r\}$ converging to $H^*(X^{S^1};\mathbb{Z}/p)$ as an algebra with

$$_{HH}E_2^{*,*} \cong HH(H^*(X;\mathbb{Z}/p)),$$

where X is a simply connected space and $HH(\ )$ denotes the Hochschild homology functor.

Note that $HH(H^*(X;\mathbb{Z}/p))$ is regarded here as a bigraded algebra with the shuffle product (see Section 8).

${}_B$**EMSS**: The bar type Eilenberg-Moore spectral sequence $\{{}_B E_r^{*,*}, d_r\}$ converging to $H^*(X^{S^1};\mathbb{Z}/p)$ as an algebra with

$$_B E_2^{*,*} \cong \mathrm{Tor}_{H^*(X;\mathbb{Z}/p) \otimes H^*(X;\mathbb{Z}/p)}(H^*(X;\mathbb{Z}/p), H^*(X;\mathbb{Z}/p)).$$

${}_C$**EMSS**: The cobar type Eilenberg-Moore spectral sequence $\{{}_C E_r^{*,*}, d_r\}$ converging to $H^*(G \times_{\mathrm{ad}} EG;\mathbb{Z}/p)$ as an algebra with

$$_C E_2^{*,*} \cong \mathrm{Cotor}_{H^*(G;\mathbb{Z}/p)}(H^*(G;\mathbb{Z}/p), \mathbb{Z}/p),$$

where the $H^*(G;\mathbb{Z}/p)$-comodule structure of $H^*(G;\mathbb{Z}/p)$ is induced by the right adjoint action $\mathrm{ad} : G \times G \to G$.

Observe that the first and second spectral sequences are of the second quadrant and the third one is of the first quadrant.

In order to construct the first spectral sequence HSS, we rely heavily on the isomorphism
$$H^*(X^{S^1}; \mathbb{Z}/p) \cong HH(C^*(X; \mathbb{Z}/p))$$
due to Jones [18], where $C^*(X; \mathbb{Z}/p)$ denotes the normalized cochain complex. Recently, Ndombol and Thomas [37] have proved that the isomorphism preserves the algebra structure under an appropriate product on $HH(C^*(X; \mathbb{Z}/p))$. This fact enables us to define an algebra structure on the HSS. For more details, we refer the reader to Section 8. On the other hand, the second spectral sequence $_B$EMSS is constructed with the isomorphism
$$\text{Tor}_{C^*(X \times X; \mathbb{Z}/p)}(C^*(X; \mathbb{Z}/p), C^*(X; \mathbb{Z}/p)) \cong H^*(X^{S^1}; \mathbb{Z}/p)$$
due to Eilenberg and Moore. Therefore we regard these two spectral sequences as essentially different from each other, although the E_2-terms of the spectral sequences are isomorphic as algebras. Observe that
$$HH(A) \cong \text{Tor}_{A \otimes A}(A, A)$$
as an algebra for any commutative algebra A.

As is well known, the cohomology algebra $H^*(BG; \mathbb{Z}/p)$ is an evenly generated polynomial algebra if and only if $H^*(G; \mathbb{Z})$ has no p-torsion. Therefore the following theorem is easily deduced from [20, Theorem 1] due to Kono and Kozima.

THEOREM 1.2. *Suppose that*
$$H^*(BG; \mathbb{Z}/p) \cong \mathbb{Z}/p[y_1, y_2, \ldots, y_l]$$
as an algebra, where the degree of the generator y_j is even for any j. Then,
$$H^*(BLG; \mathbb{Z}/p) \cong \Lambda(\bar{y}_1, \bar{y}_2, \ldots, \bar{y}_l) \otimes \mathbb{Z}/p[y_1, y_2, \ldots, y_l]$$
as an $H^*(BG; \mathbb{Z}/p)$-algebra, where $\deg \bar{y}_j = \deg y_j - 1$.

We also determine the algebra structure of $H^*(BLG; \mathbb{Z}/2)$ by using the bar type EMSS and the Steenrod operation on the spectral sequence when $H^*(BG; \mathbb{Z}/2)$ is a polynomial algebra even if $H^*(G; \mathbb{Z})$ has 2-torsion.

THEOREM 1.3. ([28, Theorem 1.6], [21, Theorem 3.1]) *Suppose that*
$$H^*(BG; \mathbb{Z}/2) \cong \mathbb{Z}/2[y_1, y_2, \ldots, y_l]$$
as an algebra in which generators of odd degree are allowed. Then
$$H^*(BLG; \mathbb{Z}/2) \cong \mathbb{Z}/2[\bar{y}_1, \ldots, \bar{y}_l] \otimes \mathbb{Z}/2[y_1, \ldots, y_l]/(\bar{y}_i^2 + \mathcal{D}Sq^{\deg y_i - 1} y_i \,;\, 1 \leq i \leq l)$$

as an $H^*(BG;\mathbb{Z}/2)$-algebra, where $\deg \bar{y}_i = \deg y_i - 1$ and

$$\mathcal{D} : H^*(BG;\mathbb{Z}/2) \longrightarrow H^{*-1}(BLG;\mathbb{Z}/2)$$

is the module derivation of degree -1 defined by $\mathcal{D}y_i = \bar{y}_i$.

By applying Theorem 1.3, we have

THEOREM 1.4. (i) As an $H^*(BSO(n);\mathbb{Z}/2) = \mathbb{Z}/2[w_2,\ldots,w_n]$-algebra,

$$H^*(BLSO(n);\mathbb{Z}/2) \cong \mathbb{Z}/2[\bar{w}_2,\ldots,\bar{w}_n] \otimes$$

$$\mathbb{Z}/2[w_2,\ldots,w_n] \Big/ \Big(\bar{w}_k^2 + \bar{w}_{2k-1} + \sum_{2 \leq i \leq k-1} (\bar{w}_{2k-i-1}w_i + w_{2k-i-1}\bar{w}_i) \Big).$$

(ii) As an $H^*(BG_2;\mathbb{Z}/2) = \mathbb{Z}/2[v_4,v_6,v_7]$-algebra,

$$H^*(BLG_2;\mathbb{Z}/2) \cong \mathbb{Z}/2[x_3,x_5] \otimes \mathbb{Z}/2[v_4,v_6,v_7] \Big/ \begin{pmatrix} x_3^4 + x_5v_7 + v_6x_3^2, \\ x_5^2 + x_3v_7 + v_4x_3^2 \end{pmatrix}.$$

(iii) As an $H^*(BF_4;\mathbb{Z}/2) = \mathbb{Z}/2[v_4,v_6,v_7,v_{16},v_{24}]$-algebra,

$$H^*(BLF_4;\mathbb{Z}/2) \cong \mathbb{Z}/2[x_3,x_5,x_{15},x_{23}] \otimes$$

$$\mathbb{Z}/2[v_4,v_6,v_7,v_{16},v_{24}] \Big/ \begin{pmatrix} x_5^2 + x_3v_7 + x_3^2v_4, \\ x_3^4 + x_5v_7 + x_3^2v_6, \\ x_{15}^2 + x_3^2v_{24} + x_{23}v_7, \\ x_{23}^2 + x_3^2v_{16}v_{24} + v_7x_{15}v_{24} + v_7v_{16}x_{23} \end{pmatrix}.$$

THEOREM 1.5. (i) As an $H^*(BSpin(7);\mathbb{Z}/2) = \mathbb{Z}/2[y_4,y_6,y_7,u_8]$-algebra,

$$H^*(BLSpin(7);\mathbb{Z}/2) \cong \mathbb{Z}/2[x_3,x_5,z_7] \otimes$$

$$\mathbb{Z}/2[y_4,y_6,y_7,u_8] \Big/ \begin{pmatrix} x_5^2 + x_3y_7 + x_3^2y_4, \\ x_3^4 + x_5y_7 + x_3^2y_6, \\ z_7^2 + x_3^2u_8 + z_7y_7 \end{pmatrix}.$$

(ii) As an $H^*(BSpin(8);\mathbb{Z}/2) = \mathbb{Z}/2[y_4,y_6,y_7,y_8,u_8]$-algebra,

$$H^*(BLSpin(8);\mathbb{Z}/2) \cong \mathbb{Z}/2[x_3,x_5,x_7,z_7] \otimes$$

$$\mathbb{Z}/2[y_4,y_6,y_7,y_8,u_8] \Big/ \begin{pmatrix} x_5^2 + x_3y_7 + x_3^2y_4, \\ x_3^4 + x_5y_7 + x_3^2y_6, \\ x_7^2 + x_3^2y_8 + x_7y_7, \\ z_7^2 + x_3^2u_8 + z_7y_7 \end{pmatrix}.$$

(iii) As an $H^*(BSpin(9);\mathbb{Z}/2) = \mathbb{Z}/2[y_4,y_6,y_7,y_8,u_{16}]$-algebra,

$$H^*(BLSpin(9);\mathbb{Z}/2) \cong \mathbb{Z}/2[x_3,x_5,x_7,z_{15}] \otimes$$

$$\mathbb{Z}/2[y_4, y_6, y_7, y_8, u_{16}] \Bigg/ \begin{pmatrix} x_5^2 + x_3y_7 + x_3^2 y_4, \\ x_3^4 + x_5 y_7 + x_3^2 y_6, \\ x_7^2 + x_3^2 y_8 + x_7 y_7, \\ z_{15}^2 + x_3^2 y_8 u_{16} + x_7 y_7 u_{16} + z_{15} y_7 y_8 \end{pmatrix}.$$

Recall that if G is a simply connected, compact simple classical Lie group, then the cohomology algebra $H^*(BG; \mathbb{Z}/p)$ is a polynomial algebra except for the case where $(G, p) = (Spin(N), 2)$ for $N \geq 10$.

When the cohomology $H^*(BG; \mathbb{Z}/p)$ is not a polynomial algebra, we apply the cobar type EMSS for investigating the structure of $H^*(BLG; \mathbb{Z}/p)$. In order to make an explicit calculation of the spectral sequence, as the first step, we need a manageable injective resolution of $\mathbb{Z}/p$ as an $H^*(G; \mathbb{Z}/p)$-comodule with a multiplication. One of the candidates for such a resolution is the twisted tensor product mentioned above. In fact, we can construct a differential graded algebra to compute the cotorsion product $\mathrm{Cotor}_{H^*(G;\mathbb{Z}/p)}(H^*(G; \mathbb{Z}/p), \mathbb{Z}/p)$ as an algebra by making use of the twisted tensor product in Theorem 1.1. It is difficult to determine explicitly the algebra structure of the cotorsion product in general although we use our differential graded algebra in the computation. In a particular case, however, the differential graded algebra enables us to determine the cotorsion product and to compute the cobar type EMSS converging to $H^*(BLG; \mathbb{Z}/p)$.

THEOREM 1.6. *As an* $H^*(BPU(3); \mathbb{Z}/3) = \mathbb{Z}/3[y_2, y_8, y_{12}] \otimes \Lambda(y_3, y_7)/(y_2 y_3, y_2 y_7, y_2 y_8 + y_3 y_7)$-*module,*

$$H^*(BLPU(3); \mathbb{Z}/3) = \mathbb{Z}/3[x_2, y_2, z_6, y_8, z_8, y_{12}] \otimes \Lambda(x_1, y_3, y_7, z_9, z_{11})/I,$$

where I is the ideal generated by elements

$x_2 y_2 + x_1 y_3,$	$y_2 y_3,$	$x_2^3,$
$y_2 z_6 + x_1 y_7,$	$x_1 z_8,$	$y_2 y_7$
$z_6 y_3 - x_2 y_7 - x_1 y_8,$	$x_1 z_9 - y_2 z_8,$	$x_2 z_8,$
$y_2 y_8 + y_3 y_7,$	$z_8 y_3 - x_1 x_2^2 z_6,$	$x_1 z_{11},$
$x_2 z_9,$	$z_9 y_3 - x_1 x_2^2 y_7,$	$x_1 z_6^2 + x_2 z_{11},$
$y_2 z_{11},$	$z_{11} y_3 + x_1 z_6 y_7,$	$z_6 z_8,$
$z_6 z_9 + z_8 y_7,$	$z_8 y_7 - x_2^2 z_{11},$	$x_2^2 z_6^2 - z_8 y_8,$
$z_9 y_7,$	$z_8^2,$	$x_2^2 z_6 y_7 + z_9 y_8,$
$z_6 z_{11} + x_1 x_2^2 y_{12},$	$z_8 z_9,$	$z_6^3,$
$z_{11} y_7 + x_1 x_2 y_3 y_{12},$	$-z_6^2 y_7 + z_{11} y_8 + x_2^2 y_3 y_{12},$	$z_8 z_{11},$

$z_9 z_{11}$.

THEOREM 1.7. *As an* $H^*(BSpin(10); \mathbb{Z}/2) = \mathbb{Z}/2[y_4, y_6, y_7, y_8, y_{10}, y_{32}]/(y_7 y_{10})$-*module,*

$$H^*(BLSpin(10); \mathbb{Z}/2) \cong \mathrm{Cotor}_{H^*(Spin(10); \mathbb{Z}/2)}(H^*(Spin(10); \mathbb{Z}/2), \mathbb{Z}/2)$$

$$\cong \mathbb{Z}/2[x_3, y_4, y_6, y_7, y_8, y_{10}, y_{32}]$$

$$\otimes \Lambda(x_5, x_7, x_9, z_{30}, z_{31}, w_{31})/I,$$

where I is the ideal generated by elements

$$x_3^4, \quad y_7 y_{10}, \quad w_{31} y_{10}, \quad z_{31} y_7, \quad x_3^2 z_{30}, \quad x_3^2 z_{31}, \quad x_9 y_7 + x_3^2 y_{10},$$

$$x_9 z_{30}, \quad x_9 w_{31}, \quad z_{30} z_{31}, \quad z_{30} w_{31}, \quad z_{31} w_{31}, \quad x_3^2 w_{31} + z_{30} y_7, \quad x_9 z_{31} + z_{30} y_{10}.$$

As mentioned above, cohomology algebra $H^*(BSpin(N); \mathbb{Z}/2)$ is not a polynomial algebra for $N \geq 10$. Therefore, it is worthwhile to compute as *the first example* to which Theorem 1.3 is not applicable.

For the rest of Introduction, $\{_{HH}E_r^{*,*}, d_r\}$ and $\{_C E_r^{*,*}, d_r\}$ denote the HSS and the cobar type EMSS respectively converging to the same cohomology algebra $H^*(BLSpin(10); \mathbb{Z}/2)$.

Let us here describe the outline of the proof of Theorem 1.7, because the proof itself is of our interest. What we pay attention to in the proof is to exchange information on the triviality of spectral sequences between $\{_{HH}E_r^{*,*}, d_r\}$ and $\{_C E_r^{*,*}, d_r\}$. The procedure is stated as follows. Let A be the cohomology Hopf algebra $H^*(Spin(10); \mathbb{Z}/2)$. First we calculate exactly the cotorsion product $\mathrm{Cotor}_A(A, \mathbb{Z}/2)$ by using the twisted tensor product associated with A, which we shall construct later in this paper. Consequently, we see that the E_2-term $_C E_2^{*,*}$ is generated by elements with total degree less than or equal to 32. A partial calculation of the Hochschild homology $HH(H^*(BSpin(10); \mathbb{Z}/2))$, which is the E_2-term of the spectral sequence $\{_{HH}E_r^{*,*}, d_r\}$, allows us to compare $_{HH}E_2^{*,*}$ with $_C E_2^{*,*}$ as a vector space up to total degrees 45. In consequence, we have the following key lemma to prove Theorem 1.7.

LEMMA 1.8. *For any integer $j \leq 45$, the HSS*

$$E_2 = HH(H^*(BSpin(10); \mathbb{Z}/2)) \implies H^*(BLSpin(10); \mathbb{Z}/2)$$

collapses at the E_2-term for total degree below j if and only if so does the cobar type EMSS

$$E_2 = \mathrm{Cotor}_{H^*(Spin(10); \mathbb{Z}/2)}(H^*(Spin(10); \mathbb{Z}/2), \mathbb{Z}/2) \implies H^*(BLSpin(10); \mathbb{Z}/2).$$

In particular, if the HSS collapses at the E_2-term for total degree ≤ 32, then the cobar type EMSS collapses at the E_2-term.

For dimensional reasons, it follows that the HSS $\{_{HH}E_r^{*,*}, d_r\}$ collapses at the E_2-term for total degree ≤ 29. By virtue of Lemma 1.8, we see that so does the cobar type EMSS $\{_CE_r^{*,*}, d_r\}$. From the knowledge of the Steenrod operation on $\{_CE_r^{*,*}, d_r\}$ and the algebra structure of $\text{Cotor}_A(A, \mathbb{Z}/2)$, and further, by comparing the E_1-term with that of the cobar type EMSS converging to $H^*(BLE_6; \mathbb{Z}/2)$, we can deduce that an algebra generator of $_CE_2^{0,30}$ is a permanent cycle. It follows from Lemma 1.8 that the HSS $\{_{HH}E_r^{*,*}, d_r\}$ also collapses at the E_2-term for total degree ≤ 30. We find just two algebra generators of $_{HH}E_2^{*,*}$ with total degree 31. One is in $_{HH}E_2^{-1,32}$ and the other is in $_{HH}E_2^{-3,34}$. It is immediate to see that the generator in $_{HH}E_2^{-1,32}$ is a permanent cycle. If the generator in $_{HH}E_2^{-3,34}$, say z, is also a permanent cycle, then we see that the HSS collapses at the E_2-term for total degree ≤ 32, because the only generator with total degree 32 is in $_{HH}E_2^{0,32}$. Lemma 1.8 implies that the cobar EMSS collapses at the E_2-term. We thus obtain Theorem 1.7. In order to prove that the element z in $_{HH}E_2^{-3,34}$ is a permanent cycle, we take the TV-model

$$\alpha : (TV_{BSpin(10)}, d) \xrightarrow{\simeq} C^*(BSpin(10); \mathbb{Z}/2)$$

which is a quasi-isomorphism; that is, the map induces an isomorphism on the cohomology. We compare the HSS $\{_{HH}\tilde{E}_r^{*,*}, \tilde{d}_r\}$ converging to $HH(TV_{BSpin(10)})$ with $\{_{HH}E_r^{*,*}, d_r\}$ converging

$$HH(C^*(BSpin(10); \mathbb{Z}/2)) \cong H^*(BLSpin(10); \mathbb{Z}/2)$$

by the isomorphism of spectral sequences induced by α. Observe that the differential of $\{_{HH}\tilde{E}_r^{*,*}, \tilde{d}_r\}$ is dominated by the differential d of the TV-model for $BSpin(10)$ (see [**25**, Lemma 2.1]). By analyzing the TV-model, we can conclude that the algebra generator $\tilde{z}$ in $_{HH}\tilde{E}_2^{-3,34}$, which corresponds to the element z, is a permanent cycle, and hence so is z.

As for the TV-model (TV_X, d) for a simply connected space X, it follows from [**10**] that the vector space V_X is isomorphic to the suspension of cohomology $H^*(\Omega X; \mathbb{Z}/p)$ and the quadratic part of the differential d can be identified with the coproduct of $H^*(\Omega X; \mathbb{Z}/p)$. In general, it is by no means easy to determine the *higher part* of the differential d of the TV-model. Even in the TV-model $(TV_{BSpin(10)}, d)$, we can not deduce an exact form of the differentials $\tilde{d}_r$ of $\{_{HH}\tilde{E}_r, \tilde{d}_r\}$. To this end, such incomplete information on the TV-model does not work well in determining differentials on $_{HH}\tilde{E}_r^{*,*}$ with total degree 30, although, as mentioned above, it is possible to conclude that the differential $\tilde{d}_r$ on $_{HH}\tilde{E}_r^{*,*}$

with total degree 31 is trivial. Fortunately, we obtain an explicit form of an important differential of $(TV_{BSpin(10)}, d)$ from the triviality of the cobar type EMSS $\{_C E_r^{*,*}, d_r\}$ (see Theorem 1.7). This fact leads us to an application of the TV-model.

Let $\mathbb{T}$ be the circle group and $\alpha : \mathbb{T} \times BLG \to BLG$ the circle action on BLG which is induced by the $\mathbb{T}$-action on the loop group LG defined by $(t\gamma)(s) = \gamma(ts)$ for $t \in \mathbb{T}$ and $\gamma \in LG$. We define the map

$$\lambda : H^*(BLG; \mathbb{Z}/p) \longrightarrow H^*(BLG; \mathbb{Z}/p)$$

of degree -1 by $\int_{S^1} \circ \alpha^*$, where $\int_{S^1}$ is the integration along the circle. Observe that λ is a derivation. A Hochschild homological interpretation of the map λ due to Jones [18] enables us to deduce the following proposition.

PROPOSITION 1.9. *There exists a decreasing filtration*

$$F^* = \{F^i H^*(BLG; \mathbb{Z}/p) \, ; \, i \leq 0\}$$

of $H^*(BLG; \mathbb{Z}/p)$ such that the map λ decreases the filtration degree by 1:

$$\lambda : F^i H^{i+j}(BLG; \mathbb{Z}/p) \longrightarrow F^{i-1} H^{i-1+j}(BLG; \mathbb{Z}/p).$$

From the knowledge about the differential of the TV-model for $BSpin(10)$, we have a result concerning the map λ in the case $G = Spin(10)$.

THEOREM 1.10. *There exist algebra generators w_i, $\bar{w}_i$ ($i = 4, 6, 7, 8, 10, 32$), ξ_{30} and ξ_{31}, which are in the filtrations F^0, F^{-1}, F^{-3} and F^{-4} of $H^*(BLSpin(10); \mathbb{Z}/2)$, respectively, such that $\lambda(w_i) = \bar{w}_i$, $\lambda(\bar{w}_i) = \lambda(\xi_{31}) = 0$ and $\lambda(\xi_{30})$ is in the filtration F^{-3}.*

The paper is organized as follows. In Section 2 we calculate the mod 2 cohomology of BLG for $G = SO(n)$. In Section 3 we also calculate the mod 2 cohomology of BLG for $G = Spin(n)$ where $7 \leq n \leq 9$. In Section 4 we calculate those of G_2 and F_4 in a similar way. In Section 5, we first prove that the twisted tensor product for a Hopf algebra has a DGA structure under an appropriate multiplication if it is well-defined. After that, it is shown that the multiplication on each twisted tensor product constructed by Kono, Mimura, Sambe and Shimada is well-defined. Consequently, we have Theorem 1.1. Exact forms of the complexes for computing the cotorsion products $\text{Cotor}_{H^*(G;\mathbb{Z}/p)}(H^*(G;\mathbb{Z}/p), \mathbb{Z}/p)$ are listed in the end of the section. Section 6 is devoted to constructing a twisted tensor product associated with the Hopf algebra $H^*(Spin(N); \mathbb{Z}/2)$. Explicit form of the complex for computing the cotorsion product for $G = Spin(10)$ is given. The purpose of the next two sections is to introduce important tools, on which we rely to prove theorems stated above; we describe, in Section 7, a "manner" for determining the homology of a given differential graded algebra as an algebra, which sums up many elaborated computations performed by several authors, and in Section 8, after recalling the notion of a strongly homotopy commutative (*shc* for short) algebra introduced by Munkholm [36], we construct the algebraic Hochschild spectral sequence, which allows us to obtain the Hochschild spectral sequence mentioned above. In Section 9, we calculate the cotorsion product $\text{Cotor}_{H^*(PU(3);\mathbb{Z}/3)}(H^*(PU(3);\mathbb{Z}/3),\mathbb{Z}/3)$ as an algebra and show that the Eilenberg-Moore spectral sequence converging to $H^*(BLPU(3);\mathbb{Z}/3)$ collapses. In Section 10, we calculate completely the cotorsion product $\text{Cotor}_{H^*(Spin(10);\mathbb{Z}/2)}(H^*(Spin(10);\mathbb{Z}/2),\mathbb{Z}/2)$ as an algebra and determine partially the algebra structure of the Hochschild homology $HH(H^*(BSpin(10);\mathbb{Z}/2))$ according to the manner stated in Section 7. Lemma 1.8 is also proved in this section. In Section 11, we not only prove Theorem 1.7 but also analyze the TV-model for $BSpin(N)$ simultaneously. Section 12 consists of the proofs of Proposition 1.9 and Theorem 1.10.

We conclude Introduction with a comment. As will be seen, our application of the notions of a twisted tensor product and of TV-models is far from a general theory. However, the novelty here is not only in the explicit calculation of the cohomology $H^*(BLSpin(10);\mathbb{Z}/2)$ but also in the new usage of TV-models combining with spectral sequences. The authors believe firmly that the manner of such explicit calculation becomes to be a seed to develop a theory of algebraic models for spaces.

2. The mod 2 cohomology of $BLSO(n)$

Let T be a maximal torus of a Lie group G, $W_G = N(T)/T$ its Weyl group and i the inclusion map $T \to G$. The image of

$$Bi^* : H^*(BG;\mathbb{Z}/p) \to H^*(BT;\mathbb{Z}/p)$$

is included in the ring of invariant forms $H^*(BT;\mathbb{Z}/p)^{W_G}$. When p is an odd prime and G is a Lie group whose integral homology is p-torsion free, the mod p cohomology of BG is isomorphic to the ring of invariant forms of its Weyl group.

For the Lie groups $SO(n)$, the mod 2 cohomology of BG can be represented as the subalgebra of the cohomology of the classifying space of the maximal elementary abelian 2-group of G in a similar way.

Let V be the maximal elementary abelian 2-group of $SO(n)$ consisting of the diagonal matrices. The mod 2 cohomology of BV is

$$H^*(BV;\mathbb{Z}/2) = \mathbb{Z}/2[t_1,\ldots,t_n]/\sigma_1,$$

where $\deg t_i = 1$ and $\sigma_1 = t_1 + \ldots + t_n$. The mod 2 cohomology of $BSO(n)$ is isomorphic to the ring of invariant forms of the symmetric group Σ_n, that is,

$$H^*(BSO(n); \mathbb{Z}/2) = H^*(BV; \mathbb{Z}/2)^{\Sigma_n} = \mathbb{Z}/2[w_2, w_3, \ldots, w_n],$$

where w_i is called the ith Stiefel-Whitney class. As is well known, the action of the squaring operation on the Stiefel-Whitney classes is given by the Wu formula:

$$(2.1) \qquad Sq^j w_k = \sum_{i=0}^{j} \binom{k-i-1}{j-i} w_{k+j-i} w_i \quad \text{for } 0 \le j \le k.$$

Applying Theorem 1.3, we obtain Theorem 1.4 (i); the details of computation are left to the reader.

3. The mod 2 cohomology of BLG for $G = Spin(n)$ ($7 \le n \le 9$)

Borel calculated in [2] the mod 2 cohomology of $BSpin(n)$ for $n \le 10$. In particular, he showed that they are polynomial algebras for $7 \le n \le 9$:

$$H^*(BSpin(7); \mathbb{Z}/2) = \mathbb{Z}/2[y_4, y_6, y_7, u_8],$$
$$H^*(BSpin(8); \mathbb{Z}/2) = \mathbb{Z}/2[y_4, y_6, y_7, y_8, u_8],$$
$$H^*(BSpin(9); \mathbb{Z}/2) = \mathbb{Z}/2[y_4, y_6, y_7, y_8, u_{16}].$$

In order to calculate the cohomology of BLG for $G = Spin(7)$, $Spin(8)$ and $Spin(9)$ by applying Theorem 1.3, we need to know the algebra structure of $H^*(BG; \mathbb{Z}/2)$ over the Steenrod algebra. Let us recall Quillen's result [39] in which he calculated the mod 2 cohomology of $BSpin(n)$; let J be the ideal of $H^*(BSO(n); \mathbb{Z}/2)$ generated by

$$w_2, Sq^1 w_2, \ldots, Sq^{2^{h-2}} \cdots Sq^1 w_2,$$

where

$$h = \begin{cases} 4l & \text{if } n = 8l+1 \\ 4l+1 & \text{if } n = 8l+2 \\ 4l+2 & \text{if } 8l+3 \le n \le 8l+4 \\ 4l+3 & \text{if } 8l+5 \le n \le 8l+8. \end{cases}$$

Let π be the natural projection

$$\pi : Spin(n) \longrightarrow SO(n).$$

We denote by y_i the image of the homomorphism $B\pi^*$ of the ith Stiefel-Whitney class.

3. THE MOD 2 COHOMOLOGY OF BLG FOR $G = Spin(n)$ ($7 \leq n \leq 9$)

THEOREM 3.1. (Quillen [39]) *The algebra structure of the mod 2 cohomology of $BSpin(n)$ is given by*

$$H^*(BSpin(n); \mathbb{Z}/2) \cong H^*(BSO(n); \mathbb{Z}/2)/J \otimes \mathbb{Z}/2[u_{2^h}],$$

where u_{2^h} is the 2^h-th Stiefel-Whitney class of the spin representation. The non-zero Stiefel-Whitney classes of the spin representation are those of degrees 2^h and $2^h - 2^i$ for $r \leq i \leq h$, where

$$r = \begin{cases} 0 & n \equiv 0, 1, 7 \mod 8 \\ 1 & n \equiv 2, 6 \mod 8 \\ 2 & n \equiv 3, 4, 5 \mod 8. \end{cases}$$

It follows from Theorem 3.1 that the total Stiefel-Whitney class of the spin representation Δ_7 of $Spin(7)$ is given as follows:

$$w(\Delta_7) = 1 + y_4 + y_6 + y_7 + u_8.$$

Using the Wu formula, the Steenrod algebra action on $H^*(BSpin(7); \mathbb{Z}/2)$ is given by the following table:

	Sq^1	Sq^2	Sq^4	Sq^8
y_4	0	y_6	y_4^2	0
y_6	y_7	0	$y_4 y_6$	0
y_7	0	0	$y_4 y_7$	0
u_8	0	0	$y_4 u_8$	u_8^2

Applying Theorem 1.3, we obtain Theorem 1.5 (i); the details of computation are left to the reader.

As is well known, $Spin(8)$ has two spin representations Δ_8^+ and Δ_8^-. We put $u_8 = w_8(\Delta_8^+)$. Then it follows from Theorem 3.1 that the total Stiefel-Whitney class of the spin representation Δ_8^+ is given as follows:

$$w(\Delta_8^+) = 1 + y_4 + y_6 + y_7 + u_8.$$

Using the Wu formula, the Steenrod algebra action on $H^*(BSpin(8); \mathbb{Z}/2)$ is given by the following table:

	Sq^1	Sq^2	Sq^4	Sq^8
y_4	0	y_6	y_4^2	0
y_6	y_7	0	$y_4 y_6$	0
y_7	0	0	$y_4 y_7$	0
y_8	0	0	$y_4 y_8$	y_8^2
u_8	0	0	$y_4 u_8$	u_8^2

Applying Theorem 1.3, we obtain Theorem 1.5 (ii); the details of computation are left to the reader.

Before considering the case $H^*(Spin(9); \mathbb{Z}/2)$, it is necessary to calculate the total Stiefel-Whitney class $w(\Delta_8^-)$. According to [1], the outer automorphism $Out(Spin(8))$ of $Spin(8)$ is isomorphic to the symmetric group Σ_3 of degree 3 which acts on the set of the representations π, Δ_8^+ and Δ_8^-, where π is the natural projection $Spin(8) \to SO(8)$. Then there is an automorphism $\sigma : Spin(8) \to Spin(8)$ such that $\sigma^*(\pi) = \Delta_8^-$, $\sigma^*(\Delta_8^+) = \pi$ and $\sigma^*(\Delta_8^-) = \Delta_8^+$. We put

$$w(\Delta_8^-) = 1 + y_4 + y_6 + y_7 + a_1 y_8 + a_2 u_8 + a_3 y_4^2,$$

where $a_i \in \mathbb{Z}/2$. Then we have

$$w_8(\Delta_8^+) = (\sigma^*)^2 w_8(\pi) = \sigma^* w_8(\Delta_8^-) = a_1(a_1 y_8 + a_2 u_8 + a_3 y_4^2) + a_2 y_8 + a_3 y_4^2.$$

Thus we obtain $a_1 = a_2 = 1$. Comparing the coefficients of the Wu formula $Sq^4 w_8 = w_4 w_8$, we obtain $a_3 = 0$.

Let $f : Spin(8) \to Spin(9)$ be the natural inclusion map. It is well known that the induced representation $f^* \Delta_9$ is isomorphic to $\Delta_8^+ \oplus \Delta_8^-$, where Δ_9 is the spin representation of $Spin(9)$. Then the total Stiefel-Whitney class of $f^* \Delta_9$ is given as follows:

$$w(f^* \Delta_9) = w(\Delta_8^+) w(\Delta_8^-)$$
$$= 1 + (y_8 + y_4^2) + (y_6^2 + y_4 y_8) + (y_7^2 + y_6 y_8) + y_7 y_8 + (y_8 u_8 + u_8^2).$$

Since $Bf^* : H^*(BSpin(9); \mathbb{Z}/2) \to H^*(BSpin(8); \mathbb{Z}/2)$ is a monomorphism, the total Stiefel-Whitney class of the spin representation Δ_9 is given as follows:

$$w(\Delta_9) = 1 + (y_8 + y_4^2) + (y_6^2 + y_4 y_8) + (y_7^2 + y_6 y_8) + y_7 y_8 + u_{16}.$$

Using the Wu formula, the Steenrod algebra action on $H^*(BSpin(8); \mathbb{Z}/2)$ is given by the following table:

	Sq^1	Sq^2	Sq^4	Sq^8	Sq^{16}
y_4	0	y_6	y_4^2	0	0
y_6	y_7	0	$y_4 y_6$	0	0
y_7	0	0	$y_4 y_7$	0	0
y_8	0	0	$y_4 y_8$	y_8^2	0
u_{16}	0	0	0	$(y_8 + y_4^2) u_{16}$	u_{16}^2

Applying Theorem 1.3, we obtain Theorem 1.5 (iii); the details of computation are left to the reader.

4. The mod 2 cohomology of BLG for $G = G_2$, F_4

Let $\mathfrak{C}$ be the Cayley algebra and $\mathfrak{J}$ the Jordan algebra of all the 3-hermitian matrices over the Cayley algebra. The compact exceptional Lie groups G_2 and F_4 can be realized as the automorphism groups $\mathrm{Aut}(\mathfrak{C})$ and $\mathrm{Aut}(\mathfrak{J})$ respectively (see, for example [46]).

Borel showed in [2] that the mod 2 cohomology rings of BG_2 and BF_4 are polynomial algebras:

$$H^*(BG_2; \mathbb{Z}/2) = \mathbb{Z}/2[v_4, v_6, v_7],$$
$$H^*(BF_4; \mathbb{Z}/2) = \mathbb{Z}/2[v_4, v_6, v_7, v_{16}, v_{24}].$$

According to the transgression theorem, we have

$$Sq^2 v_4 = v_6, \qquad Sq^1 v_6 = v_7,$$
$$Sq^8 v_{16} \equiv v_{24} \quad \text{modulo decomposable elements.}$$

Let $\mathfrak{C}_0$ be the space consisting of the imaginary part of $\mathfrak{C}$. Since $\mathfrak{C}_0$ is invariant under the action of G_2, there is a representation

$$\rho : G_2 \longrightarrow SO(7).$$

We consider the image of the homomorphism $B\rho^*$ of the Stiefel-Whitney classes. As is well known, there is a subgroup of G_2 which is isomorphic to $SU(3)$. Since the composition

$$SU(3) \xrightarrow{f} G_2 \xrightarrow{\rho} SO(7)$$

is the natural inclusion $SU(3) \to SO(6) \to SO(7)$, we obtain

$$f^* w_i(\rho) = 0, \quad \text{for } i = 1, 2, 3, 5, 7,$$
$$f^* w_4(\rho) = c_2, \qquad f^* w_6(\rho) = c_3.$$

Then it is easy to see that

$$w_i(\rho) = 0, \quad \text{for } i = 1, 2, 3, 5,$$
$$w_4(\rho) = v_4, \qquad w_6(\rho) = v_6.$$

Since $Sq^1 w_6 = w_7$, we obtain $w_7(\rho) = v_7$. Using the Wu formula, we see that the Steenrod algebra action on $H^*(BG_2; \mathbb{Z}/2)$ is given by the following table:

	Sq^1	Sq^2	Sq^4
v_4	0	v_6	v_4^2
v_6	v_7	0	$v_4 v_6$
v_7	0	0	$v_4 v_7$

Applying Theorem 1.3, we obtain Theorem 1.4 (ii); the details of computation are left to the reader. Theorem 1.4 (ii) is shown also in [**4**].

Let $\mathfrak{J}_0$ be the subspace of $\mathfrak{J}$ consisting of matrices with trace 0. $\mathfrak{J}_0$ is invariant under the action of F_4 so that there is a representation

$$\rho : F_4 \longrightarrow SO(26).$$

We consider the image of the homomorphism $B\rho^*$ of the Stiefel-Whitney classes. Let

$$E_1 = \begin{pmatrix} 1 & 0 & 0 \\ 0 & 0 & 0 \\ 0 & 0 & 0 \end{pmatrix} \in \mathfrak{J}.$$

It is easy to see that the subgroup of F_4 fixing E_1 is isomorphic to $Spin(9)$. Thus there is a monomorphism $f : Spin(9) \to F_4$. The induced representation $f^*\rho$ is isomorphic to $\Delta_9 \oplus \pi \oplus 1$, where Δ_9 is the spin representation $Spin(9) \to SO(16)$ and π is the natural projection $Spin(9) \to SO(9)$ (see Theorem 15.1 of [**47**]). Since we have

$$w(\Delta_9) = 1 + (y_8 + y_4^2) + (y_4 y_8 + y_6^2) + (y_6 y_8 + y_7^2) + y_7 y_8 + u_{16},$$
$$w(\pi) = 1 + y_4 + y_6 + y_7 + y_8,$$

we obtain

$$f^*w(\rho) = w(\Delta_9) \cdot w(\pi)$$
$$= 1 + y_4 + y_6 + y_7 + y_4^2 + (y_6^2 + y_4^3) + (y_7^2 + y_4^2 y_6) + y_4^2 y_7$$
$$+ (u_{16} + y_8^2 + y_4 y_6^2) + (y_6^3 + y_4 y_7^2) + y_6^2 y_7 + (y_4 u_{16} + y_6 y_7^2 + y_4 y_8^2)$$
$$+ y_7^3 + (y_6 u_{16} + y_6 y_8^2) + (y_7 u_{16} + y_7 y_8^2) + y_8 u_{16}.$$

Since $f^*w_{16}(\rho)$ is an indecomposable element, so is the element $w_{16}(\rho)$. Then we can put $v_{16} = w_{16}(\rho) + v_4 v_6^2$. Since $Sq^8 w_{16} = w_{24} + w_8 w_{16}$, we can put $v_{24} = w_{24}(\rho)$. Then the homomorphism $Bf^* : H^*(BF_4; \mathbb{Z}/2) \to H^*(BSpin(9); \mathbb{Z}/2)$ is a monomorphism, since the Serre spectral sequence associated with the fibration $F_4/Spin(9) \to BSpin(9) \to BF_4$ collapses. Then it is easy to obtain the Stiefel-Whitney classes of the representation ρ:

$$w_i(\rho) = 0 \quad \text{for } i = 1,2,3,5,9,10,11,13,17,$$
$$w_i(\rho) = v_i \quad \text{for } i = 4,6,7,24,$$
$$w_8(\rho) = v_4^2,$$
$$w_{12}(\rho) = v_6^2 + v_4^3,$$
$$w_{14}(\rho) = v_7^2 + v_4^2 v_6,$$

$$w_{15}(\rho) = v_4^2 v_7,$$
$$w_{16}(\rho) = v_{16} + v_4 v_6^2,$$
$$w_{18}(\rho) = v_6^3 + v_4 v_7^2,$$
$$w_{19}(\rho) = v_6^2 v_7,$$
$$w_{20}(\rho) = v_4 v_{16} + v_6 v_7^2,$$
$$w_{21}(\rho) = v_7^3,$$
$$w_{22}(\rho) = v_6 v_{16},$$
$$w_{23}(\rho) = v_7 v_{16}.$$

Using the Wu formula, the Steenrod algebra action on $H^*(BF_4; \mathbb{Z}/2)$ is given by the following table:

	Sq^1	Sq^2	Sq^4	Sq^8	Sq^{16}
v_4	0	v_6	v_4^2	0	0
v_6	v_7	0	$v_4 v_6$	0	0
v_7	0	0	$v_4 v_7$	0	0
v_{16}	0	0	0	$v_{24} + v_4^2 v_{16}$	v_{16}^2
v_{24}	0	0	$v_4 v_{24}$	$v_4^2 v_{24}$	$v_{16} v_{24} + v_4 v_6^2 v_{24}$

Applying Theorem 1.3, we obtain Theorem 1.4 (iii); the details of computation are left to the reader.

5. A multiplication on a twisted tensor product

We begin by recalling the twisted tensor product due to Brown [3] (see also [41] or [23]). Let A be a coalgebra over $\mathbb{Z}/p$ with coproduct ϕ_A and augmentation ε. Let L be a $\mathbb{Z}/p$-subspace of A, $\iota : L \to A$ the inclusion and $\theta : A \to L$ a map such that $\theta \circ \iota = \mathrm{id}_L$. We define a map $\bar{\theta} : A \to sL$ by $\bar{\theta} = s \circ \theta$ and $\bar{\iota} : sL \to A$ by $\bar{\iota} = \iota \circ s^{-1}$, where $s : L \to sL$ is the suspension. Construct the tensor product $X = T(sL)$ and denote by ψ the product in $T(sL)$. The map $\bar{\theta}$ induces a map $A \to T(sL)$ which is again denoted by $\bar{\theta}$. Let I be the ideal of $T(sL)$ generated by $(\psi \circ (\bar{\theta} \otimes \bar{\theta})) \circ \phi_A)(\mathrm{Ker}\,\bar{\theta})$. The twisted tensor product (W, d) with respect to $\bar{\theta}$ is defined as follows. We put

$$W = A \otimes X/I = A \otimes \overline{X}$$

and define the differential operator d_W by

$$d_W = 1 \otimes d_{\overline{X}} + (1 \otimes \psi) \circ (1 \otimes \bar{\theta} \otimes 1) \circ (\phi_A \otimes 1),$$

where
$$d_{\overline{X}} = -\psi \circ (\bar{\theta} \otimes \bar{\theta}) \circ \phi_A \circ \bar{\iota}.$$

Observe that $\overline{X}$ itself is a differential graded algebra with $d_{\overline{X}}$ as the differential. We may denote the twisted tensor product W with respect to $\bar{\theta} : A \to sL$ by $A \otimes_\theta \overline{X}$. The element $\bar{\theta}x$ in X or $\overline{X}$ is denoted by θx.

Let G be a compact, simply connected, simple exceptional Lie group. Then it is known [**31**] that a suitable choice of a subspace L of $H^*(G; \mathbb{Z}/p)$ makes the twisted tensor product into an injective resolution
$$0 \longrightarrow \mathbb{Z}/p \longrightarrow H^*(G; \mathbb{Z}/p) \otimes_\theta \overline{X}$$
over the coalgebra $H^*(G; \mathbb{Z}/p)$. Moreover the algebra structure of $\overline{X}$ induces that of the complex
$$(\mathbb{Z}/p \square_{H^*(G;\mathbb{Z}/p)}(H^*(G; \mathbb{Z}/p) \otimes_\theta \overline{X}), 1 \square d_W) = (\overline{X}, d_{\overline{X}}).$$
Consequently we have, as an algebra,
$$\mathrm{Cotor}_{H^*(G;\mathbb{Z}/p)}(\mathbb{Z}/p, \mathbb{Z}/p) \cong H(\overline{X}, d_{\overline{X}}).$$

In this section, we define a multiplication m_W on the twisted tensor product $A \otimes_\theta \overline{X}$ for a Hopf algebra A such that the differential d_W is derivative with respect to the multiplication. More precisely, we prove the following theorem.

THEOREM 5.1. *Let A be a Hopf algebra over $\mathbb{Z}/p$. For any elements $a \otimes \theta x$ and $b \otimes \theta y$ of $A \otimes_\theta \overline{X}$, define $m_W : A \otimes_\theta \overline{X} \otimes A \otimes_\theta \overline{X} \to A \otimes_\theta \overline{X}$ by*

$$(*) \quad m_W(a \otimes \theta x \otimes b \otimes \theta y) = a \otimes \theta x \cdot b \otimes \theta y = \sum_i (-1)^{|\theta x||b'_i|} ab'_i \otimes \theta(xb''_i)\theta y,$$

and
$$(\theta x_1 \cdots \theta x_s) \cdot a = (\theta x_1(\theta x_2(\cdots (\theta x_s \cdot a))\cdots),$$
where $\phi_A(b) = \sum_i b'_i \otimes b''_i$. If m_W is well-defined, then $(A \otimes_\theta \overline{X}, d_W, m_W)$ is a differential graded algebra. In particular, $(A \otimes_\theta \overline{X}, d_W, m_W)$ is the twisted tensor product of differential graded algebras $(A, 0)$ and $(\overline{X}, d_X)$ in the sense of Hess.

In order to verify that the multiplication m_W is well-defined, we will assume that the $\mathbb{Z}/p$-subspace L of A satisfies the following condition:

(I) *There exist the set $Q = QA$ of indecomposable elements of A and a basis $\{x_i\}$ of L such that $\{x_i\} \subset Q \cup Q^2$, where $Q^2 = \{\alpha^2 | \alpha \in Q \cap \mathrm{Prim}\, A\}$ and, as an algebra,*
$$A \cong \bigotimes_{x_s \in S} \mathbb{Z}/p[x_s]/(x_s^{p^{n_s}}) \otimes \bigotimes_{x_t \in T} \Lambda(x_j),$$
where $S \cup T = Q \cap \{x_i\}$ and $S \cap T = \emptyset$.

We assume further that

(II) $(\psi \circ (\bar{\theta} \otimes \bar{\theta}) \circ \phi_A)(\mathrm{Ker}\,\bar{\theta}) = \mathbb{Z}/p\{(\psi \circ (\bar{\theta} \otimes \bar{\theta}) \circ \phi_A)(x_i x_j) | x_i, x_j \in \{x_i\}, i \neq j\}$,

(III) *for any* $a \in Q$, $\theta(y a_i'') = 0$ *for any* $y \in \bar{A}$, *where* $\phi_A(a) = \sum_i a_i' \otimes a_i'' + a \otimes 1 + 1 \otimes a$

and that

(IV) *for any* x *and* $y \in \{x_i\}$, $\theta(xy) \neq 0$ *if and only if* $x = y$ *and* $x^2 \in Q^2$.

We remark here that the conditions (I), (II) (III) and (IV) hold in the cases $(PU(3), 3)$, $(F_4, 3)$, $(E_8, 3)$, (E_6, p), (E_7, p) for $p = 2$ and 3 which have been studied by Kono, Mimura, Sambe and Shimada ([**23**], [**24**], [**32**], [**33**]).

THEOREM 5.2. *If $p = 2$ or 3 and if the conditions* (I), (II), (III) *and* (IV) *hold, then the multiplication m_W is well-defined.*

In the case where $A = H^*(E_8; \mathbb{Z}/5)$, explicit calculation for the differential d_W and the multiplication m_W on $A \otimes_\theta \overline{X}$ allows us to obtain the following theorem.

THEOREM 5.3. *Let $A \otimes_\theta \overline{X}$ be the twisted tensor product of $H^*(E_8; \mathbb{Z}/5)$ constructed in [**34**]. Then $(A \otimes_\theta \overline{X}, d_W, m_W)$ is a well-defined differential graded algebra.*

In the case where $A = H^*(E_8; \mathbb{Z}/2)$, indecomposable elements x on A can be chosen so that $\bar{\phi}(x)$ is in $P \otimes P$, where P is the $\mathbb{Z}/2$-subspace of A consisting of primitive elements. Thanks to this fact, we can easily verify that the multiplication m_W is well-defined.

THEOREM 5.4. *Let $A \otimes_\theta \overline{X}$ be the twisted tensor product associated with $H^*(E_8; \mathbb{Z}/2)$ constructed in [**31**]. Then $(A \otimes_\theta \overline{X}, d_W, m_W)$ is a well-defined differential graded algebra.*

Now we recall the definition of the twisted tensor product due to Hess [**15**].

DEFINITION 5.5. A twisted tensor product of two differential graded algebras (A, d) and (B, e), denoted by $(A \tilde{\otimes} B, D)$, is a differential graded algebra which satisfies the following conditions:

(1) as a graded R-module $A \tilde{\otimes} B \approx A \otimes B$;

(2) for all $a \in A$ and $b \in B$,

$$(1 \otimes b) \odot (a \otimes 1) - (-1)^{\deg a \deg b} a \otimes b \in A^{<\deg a} \otimes B^+$$

and $(a \otimes 1) \odot (1 \otimes b) = a \otimes b$, where $\odot$ denotes the product in $A \tilde{\otimes} B$;

(3) The sequence
$$0 \longrightarrow (B,e) \xrightarrow{\iota} (A\tilde{\otimes}B, D) \xrightarrow{\pi} (A,d) \longrightarrow 0$$
is a sequence of DGA-morphisms, exact at (A,d) and (B,e), where $\iota(b) = 1 \otimes b$ and $\pi(a \otimes b) = \varepsilon(b)a$ with ε the augmentation of B.

PROOF OF THEOREM 1.1. From Theorems 5.1, 5.2, 5.3 and 5.4, it follows that the product defined in Theorem 5.1 makes the twisted tensor product into the one in the sense of Hess (see [**15**, Definition, p. 33]). □

The first step in computing the cobar type EMSS $\{_C E_r, d_r\}$ is to clarify the algebra structure of $\mathrm{Cotor}_{H^*(G;\mathbb{Z}/p)}(H^*(G;\mathbb{Z}/p), \mathbb{Z}/p)$. To this end, we employ the differential graded algebra $(A \otimes_\theta \overline{X}, d_W, m_W)$, which is an injective resolution as well. In order to prove that the multiplication m_W induces the algebra structure on $\mathrm{Cotor}_A(A, \mathbb{Z}/p)$, it suffices to prove

PROPOSITION 5.6. *Let* $\mu : A \otimes A \to A$ *be the multiplication of* A. *Then the map*
$$m_W : A \otimes_\theta \overline{X} \otimes A \otimes_\theta \overline{X} \longrightarrow A \otimes_\theta \overline{X}$$
is a μ-morphism if m_W is well-defined, that is, the following diagram is commutative:

$$\begin{array}{ccc}
A \otimes_\theta \overline{X} \otimes A \otimes_\theta \overline{X} & \xrightarrow{\psi_1} & (A \otimes A) \otimes A \otimes_\theta \overline{X} \otimes A \otimes_\theta \overline{X} \\
{\scriptstyle m_W}\downarrow & & \downarrow{\scriptstyle \mu \otimes m_W} \\
A \otimes_\theta \overline{X} & \xrightarrow{\psi_2} & A \otimes A \otimes_\theta \overline{X},
\end{array}$$

where ψ_1 and ψ_2 are the comodule structures of $A \otimes_\theta \overline{X} \otimes A \otimes_\theta \bar{X}$ and $A \otimes_\theta \overline{X}$ respectively.

Let A be the Hopf algebra $H^*(G; \mathbb{Z}/p)$. Since
$$\mathrm{ad}^* \otimes 1 : A \otimes \overline{X} \longrightarrow A \square_A (A \otimes \overline{X})$$
is an isomorphism with the inverse $1 \otimes \varepsilon \otimes 1$, we can define a differential d on $A \otimes \overline{X}$ by the compositions
$$A \otimes \overline{X} \xrightarrow{\mathrm{ad}^* \otimes 1} A \square_A (A \otimes \overline{X}) \xrightarrow{\mathrm{inc.}} A \otimes (A \otimes \overline{X}) \xrightarrow{1 \otimes d_W} A \otimes (A \otimes \overline{X}) \xrightarrow{1 \otimes \varepsilon \otimes 1} A \otimes \overline{X}.$$

A straightforward calculation for the differential $d : A \otimes \overline{X} \to A \otimes \overline{X}$ enables us to obtain the following explicit formula for d.

5. A MULTIPLICATION ON A TWISTED TENSOR PRODUCT

LEMMA 5.7. *We write* $\phi_A(x) = x \otimes 1 + 1 \otimes x + \sum_i x'_i \otimes x''_i$ *for* $x \in A$. *If* x'_i *is primitive for any* i, *then*

$$dx = -\sum_i (-1)^{|x''_i|(|x'_i|+1)} x''_i \otimes \theta x'_i + \sum_i (-1)^{|x'_i|} x'_i \otimes \theta x''_i \,.$$

The multiplication m_W on the twisted tensor product $A \otimes_\theta \overline{X}$ induces a multiplication m on $A \otimes \overline{X}$ defined by

$$A \otimes \overline{X} \otimes A \otimes \overline{X} \xrightarrow{\mathrm{ad}^* \otimes 1 \otimes \mathrm{ad}^* \otimes 1} A \square_A (A \otimes \overline{X}) \otimes A \square_A (A \otimes \overline{X})$$

$$\xrightarrow{\mathrm{inc.}} A \otimes (A \otimes \overline{X}) \otimes A \otimes (A \otimes \overline{X}) \longrightarrow A \otimes A \otimes (A \otimes \overline{X}) \otimes (A \otimes \overline{X})$$

$$\xrightarrow{m_A \otimes m_W} A \otimes (A \otimes \overline{X}) \xrightarrow{1 \otimes \varepsilon \otimes 1} A \otimes \overline{X}.$$

We can obtain an explicit formula for the multiplication m on $A \otimes \overline{X}$.

LEMMA 5.8. *We write* $\phi_A(a) = a \otimes 1 + 1 \otimes a + \sum_i a'_i \otimes a''_i$ *for* $a \in A$. *If* a'_i *is primitive for any* i, *then*

$$\theta x \cdot a = (-1)^{|\theta x||a|} a \otimes \theta x - \sum_i (-1)^{|a''_i||a'_i|+|a''_i||\theta x|} a''_i \otimes \theta(xa'_i) + \sum_i (-1)^{|a'_i||\theta x|} a'_i \otimes \theta(xa''_i).$$

Thus we can obtain a differential graded algebra $(A \otimes \overline{X}, d, m)$. From the construction of this differential graded algebra, we have

THEOREM 5.9. *Let* A *be the Hopf algebra* $H^*(G; \mathbb{Z}/p)$. *If the twisted tensor product* $(A \otimes_\theta \overline{X}, d_W, m_W)$ *is a well-defined acyclic differential graded algebra, then, as an algebra,*

$$\mathrm{Cotor}_A(A, \mathbb{Z}/p) \cong H(A \otimes \overline{X}, d, m).$$

Before proving Proposition 5.6, we prepare a lemma. We define a multiplication m on $A \otimes X$ by the same fashion as in Theorem 5.1, that is, $(*)$ in the page 16 for $a \otimes \theta x, b \otimes \theta y \in A \otimes X$.

LEMMA 5.10. *The multiplication* m *on* $A \otimes X$ *is associative.*

PROOF. For any a, b and $c \in A$, we see that

(5.1)
$$\begin{aligned}
&((a \otimes \theta x_1 \cdots \theta x_n) \cdot (b \otimes \theta y_1 \cdots \theta y_m)) \cdot (c \otimes \theta z_1 \cdots \theta z_l) \\
&= (a(\theta x_1 \cdots \theta x_n \cdot b \otimes \theta y_1 \cdots \theta y_m)) \cdot (c \otimes \theta z_1 \cdots \theta z_l) \\
&= a(((\theta x_1 \cdots \theta x_n \cdot b) \theta y_1 \cdots \theta y_m) \cdot c) \theta z_1 \cdots \theta z_l \\
&= a((\theta x_1 \cdots \theta x_n \cdot b) \cdot (\theta y_1 \cdots \theta y_m \cdot c)) \theta z_1 \cdots \theta z_l
\end{aligned}$$

and that

$$
\begin{aligned}
(a \otimes \theta x_1 \cdots \theta x_n) &\cdot ((b \otimes \theta y_1 \cdots \theta y_m) \cdot (c \otimes \theta z_1 \cdots \theta z_l)) \\
&= a \otimes \theta x_1 \cdots \theta x_n \cdot ((b \otimes \theta y_1 \cdots \theta y_m \cdot c) \theta z_1 \cdots \theta z_l) \\
&= a((\theta x_1 \cdots \theta x_n) \cdot (b \cdot (\theta y_1 \cdots \theta y_m \cdot c))) \theta z_1 \cdots \theta z_l.
\end{aligned}
\tag{5.2}
$$

If the equality

$$
(z \cdot b) \cdot d = z \cdot (b \cdot d) \tag{$\star$}
$$

holds for any $b, d \in A$ and $z \in \bar{X}$, one sees that the multiplication m_W is associative on $A \otimes X$, since $\theta y_1 \cdots \theta y_m \cdot c$ can be written $\sum d_i \otimes w_i$ for some $d_i \in A$ and $w_i \in \bar{X}$.

By making use of the induction on the number of factors of z, one can prove that the equality $(\star)$ holds. Suppose that the equality

$$
(\theta x_1 \cdots \theta x_l \cdot b) \cdot d = \theta x_1 \cdots \theta x_l \cdot (b \cdot d) \tag{$\star$)$_N$}
$$

holds for $1 \leq l \leq N$. By the definition of m_W, we have

$$
((\theta x \cdot b) \theta y_1 \cdots \theta y_N) \cdot c = (\theta x \cdot b) \cdot (\theta y_1 \cdots \theta y_N \cdot c).
$$

We can write $\theta y_1 \cdots \theta y_N \cdot c = \sum d_i \otimes z_i$. Therefore we see

$$
\begin{aligned}
(\theta x \cdot b) \cdot (\theta y_1 \cdots \theta y_N \cdot c) &= (\theta x \cdot b) \cdot \sum d_i \otimes z_i \\
&= \sum ((\theta x \cdot b) \cdot d_i) z_i \\
&= \sum (\theta x \cdot (b \cdot d_i)) z_i && \text{(from $(\star)_N$)} \\
&= \sum \theta x \cdot (b \cdot d_i \otimes z_i) \\
&= \theta x \cdot (b \cdot \sum d_i \otimes z_i) \\
&= \theta x \cdot (b \cdot (\theta y_1 \cdots \theta y_N \cdot c)).
\end{aligned}
$$

Thus (5.1), (5.2) and the above equalities enable us to conclude that

$$
((a \otimes \theta x) \cdot (b \otimes \theta y_1 \cdots \theta y_N)) \cdot (c \otimes \theta z) = (a \otimes \theta x) \cdot ((b \otimes \theta y_1 \cdots \theta y_N) \cdot (c \otimes \theta z)) \tag{5.3}
$$

for any a, b and $c \in A$. Since we can write $\theta y_1 \cdots \theta y_N \cdot b$ as $\sum \tilde{b}_i \otimes \theta \tilde{y}_{1_i} \cdots \theta \tilde{y}_{N_i}$, it follows from the assumption of the induction and (5.3) that

$$
\begin{aligned}
(\theta x \theta y_1 &\cdots \theta y_N) \cdot (b \cdot d) \\
&= \theta x (\theta y_1 \cdots \theta y_N \cdot (b \cdot d)) \\
&= \theta x (\theta y_1 \cdots \theta y_N \cdot b) \cdot d &&= \theta x ((\sum \tilde{b}_i \otimes \theta \tilde{y}_{1_i} \cdots \theta \tilde{y}_{N_i}) \cdot d \\
&= \sum \theta x ((\tilde{b}_i \otimes \theta \tilde{y}_{1_i} \cdots \theta \tilde{y}_{N_i}) \cdot d) &&= \sum (\theta x \cdot (\tilde{b}_i \otimes \theta \tilde{y}_{1_i} \cdots \theta \tilde{y}_{N_i})) \cdot d \\
&= (\theta x \cdot (\theta y_1 \cdots \theta y_N \cdot b)) \cdot d &&= (\theta x \theta y_1 \cdots \theta y_N \cdot b) \cdot d.
\end{aligned}
$$

5. A MULTIPLICATION ON A TWISTED TENSOR PRODUCT

In order to start the induction, we must verify that $(\theta x \cdot b) \cdot d = \theta x \cdot (b \cdot d)$ for any $b, d \in A$. We write $\phi(b) = \sum_i b'_i \otimes b''_i$ and $\phi(d) = \sum_j d'_j \otimes d''_j$. Then we have

$$(\theta x \cdot b) \cdot d = (\sum_i (-1)^{|\theta x||b'_i|} b'_i \otimes \theta(xb''_i)) \cdot d$$
$$= \sum_{i,j} (-1)^{|\theta x||b'_i| + |\theta x b''_i||d'_j|} b'_i d'_j \otimes \theta(xb''_i d''_j)$$

and

$$\theta x \cdot (b \cdot d) = \sum_{i,j} (-1)^{|\theta x||b'_i d'_j| + |b''_i||d'_j|} b'_i d'_j \otimes \theta(xb''_i d''_j).$$

Since

$$|\theta x||b'_i| + |\theta x b''_i||d'_j| = |x||b'_i| + |b'_i| + |x||d'_j| + |b''_i||d'_j| + |d'_j|$$
$$= |x|(|b'_i| + |d'_j|) + |b'_i| + |d'_j| + |b''_i||d'_j|,$$

it follows that $(\theta x \cdot b) \cdot d = \theta x \cdot (b \cdot d)$. This completes the proof. □

PROOF OF PROPOSITION 5.6. We note that ψ_1 is defined by

$$\psi_1 = (1 \otimes T \otimes 1 \otimes 1 \otimes 1) \circ (1 \otimes 1 \otimes T \otimes 1 \otimes 1) \circ (\phi \otimes 1 \otimes \phi \otimes 1).$$

Assume that, for any $a, b \in A$ and $z \in QX \cdots QX$ (n-times), $w \in X$,

$(*)_n \qquad \psi_2(a \otimes z \cdot b \otimes w) = m \otimes m_W \circ \psi_1(a \otimes z \otimes b \otimes w).$

From Lemma 5.10, we see that the multiplication m_W on $A \otimes_\theta \overline{X}$ is also associative if m_W is well-defined. This fact allows us to see that

$$\psi_2(a \otimes z\theta x \cdot b \otimes w) = \psi_2(a \otimes z \cdot (\theta x \cdot b) \otimes w)).$$

By putting $\phi(b) = \sum_j b'_j \otimes b''_j$, $\phi(b'_j) = \sum_l \bar{b}_{jl} \otimes \bar{\bar{b}}_{jl}$ and $\phi(a) = \sum_i a'_i \otimes a''_i$, we can proceed as

$$\psi_2(a \otimes z \cdot \sum_j (-1)^{|\theta x||b'_j|} b'_j \otimes \theta(xb''_j) \cdot w))$$
$$= m \otimes m_W \circ \psi_1(\sum_j (-1)^{|\theta x||b'_j|} a \otimes z \otimes b'_j \otimes \theta(xb''_j)w) \qquad \text{(from } (*)_n)$$
$$= \sum_l \sum_{j,i} (-1)^{|\theta x||b'_j|} (-1)^{(|a''_i| + |z|)|\bar{b}_{jl}|} a'_i \bar{b}_{jl} \otimes ((a''_i \otimes z) \cdot \bar{\bar{b}}_{jl}) \theta(xb''_j) w.$$

On the other hand, we have

$$m \otimes m_W \circ \psi_1(a \otimes z\theta x \otimes b \otimes w)$$
$$= m \otimes m_W (\sum_{i,j} (-1)^{|z\theta x| + |a''_i|)|b'_j|} a'_i \otimes b'_j \otimes a'' \otimes z\theta x \otimes b''_j \otimes w)$$
$$= \sum_{i,j} (-1)^{|z\theta x| + |a''_i|)|b'_j|} a'_i b'_j \otimes (a'' \otimes z \cdot (\theta x \cdot b''_j)) \otimes w$$

$$= \sum_s \sum_{i,j} (-1)^{(|z\theta x|+|a_i''|)|b_j'|+|\theta x||\tilde{b}_{js}|} a_i' b_j' \otimes (a_i'' \otimes z \cdot (\tilde{b}_{js} \otimes \theta(x\tilde{\tilde{b}}_{js})))w,$$

where $\phi(b_j'') = \sum_s \tilde{b}_{js} \otimes \tilde{\tilde{b}}_{js}$. From the coassociativity of ϕ, it follows that

$$\sum_{j,s} b_j' \otimes \tilde{b}_{js} \otimes \tilde{\tilde{b}}_{js} = \sum_{l,j} \bar{b}_{jl} \otimes \bar{\bar{b}}_{jl} \otimes b_j''.$$

Thus we see that there exists a bijection between the set of indexes (j,l) and that of indexes (j,s) such that $b_j' = \bar{b}_{j_1 l}$, $\tilde{b}_{js} = \bar{\bar{b}}_{j_1 l}$ and $\tilde{\tilde{b}}_{js} = b_{j_1}''$ if (j,s) is mapped to (j_1, l) under the correspondence. Therefore, it is easily seen that $(-1)^{|b_j'|+|\tilde{b}_{js}|} = (-1)^{|\bar{b}_{j_1 l}|+|\bar{\bar{b}}_{j_1 l}|} = (-1)^{|b_{j_1}''|}$. The fact enables us to conclude that

$$(-1)^{(|z\theta x|+|a_i''|)|b_j'|+|\theta x||\tilde{b}_{js}|} = (-1)^{(|z|+|a_i''|)|\bar{b}_{j_1 l}|+|\theta x|(|b_j'|+|\tilde{b}_{js}|)}$$
$$= (-1)^{|\theta x||b_{j_1}'|+(|z|+|a_i''|)|\bar{b}_{j_1 l}|}.$$

Therefore it follows that the equality $(*)_{n+1}$ holds if so does $(*)_n$. The equality

$$\psi_2(a \otimes 1 \cdot b \otimes w) = m \otimes m_W \otimes \psi_1(a \otimes 1 \otimes b \otimes w)$$

holds, since A is a Hopf algebra. Thus we have proved Proposition 5.6. □

Before proving Theorem 5.1, we prepare two lemmas.

LEMMA 5.11. *If m_W is well-defined and if the following equalities* (i) *and* (ii) *hold, then the differential d on $W = A \otimes_\theta \overline{X}$ is a derivation with respect to the multiplication m_W:*
(i) $d(b \cdot a \otimes z) = d(b) \cdot a \otimes z + (-1)^{|b|} b \cdot d(a \otimes z)$ *for $z \in \overline{X}$ and $a, b \in A$,*
(ii) $d(\theta x \cdot a \otimes z) = d(\theta x) \cdot a \otimes z + (-1)^{|\theta x|} \theta x \cdot d(a \otimes z)$ *for $a \in Q$ and $z \in \overline{X}$.*

LEMMA 5.12. *Suppose that m_W is well-defined. Then the equalities* (i) *and* (ii) *in Lemma 5.11 hold.*

PROOF OF THEOREM 5.1. The result follows from Lemmas 5.11 and 5.12. □

PROOF OF LEMMA 5.11. Let (ii)$_n$ and (iii)$_n$ denote the following equalities

$$d(\theta x \cdot a \otimes z) = d(\theta x) \cdot a \otimes z + (-1)^{|\theta x|} \theta x \cdot d(a \otimes z)$$

for $a \in Q \cdots Q$ (n-times), $z \in \overline{X}$, and

$$d(c \otimes \theta x \cdot a \otimes z) = d(c \otimes \theta x) \cdot a \otimes z + (-1)^{|c|+|\theta x|} \theta x \cdot d(a \otimes z)$$

for $a \in Q \cdots Q$ (n-times), $c \in A$ and $z \in \overline{X}$ respectively. Observe that the equality (ii)$_1$ implies (ii). Suppose that (i) and (ii)$_n$ hold. Then, for $a \in Q \cdots Q$ (n-times), we have

$$d(c \otimes \theta x \cdot a \otimes z) = d(c \cdot (\theta x \cdot a \otimes z))$$

5. A MULTIPLICATION ON A TWISTED TENSOR PRODUCT

$$= d(c) \cdot (\theta x \cdot a \otimes z) + (-1)^{|c|} c \cdot d(\theta x \cdot a \otimes z) \quad \text{(from (i))}$$

$$= d(c) \cdot (\theta x \cdot a \otimes z)$$
$$+ (-1)^{|c|} c \{ d(\theta x) \cdot a \otimes z + (-1)^{|\theta x|} \theta x \cdot d(a \otimes z) \} \quad \text{(from (ii)}_n\text{)}.$$

On the other hand, it follows from (i) that

$$d(c \otimes \theta x) \cdot a \otimes z + (-1)^{|c|+|\theta x|} c \otimes \theta x \cdot d(a \otimes z)$$
$$= (d(c) \cdot \theta x + (-1)^{|c|} c d(\theta x)) \cdot a \otimes z + (-1)^{|c|+|\theta x|} c \otimes \theta x \cdot d(a \otimes z).$$

Thus we see that if (i) and (ii)$_n$ hold, then so does (iii)$_n$. Suppose that the equality (iii)$_n$ holds. Then we have, for $a \in Q$ and $b \in Q \cdots Q$ (n-times),

$$d(\theta x \cdot (ab \otimes z)) = d((\theta x \cdot a) \cdot b \otimes z))$$
$$= d(\theta x \cdot a) \cdot b \otimes z + (-1)^{|\theta x|+|a|} (\theta x \cdot a) d(b \otimes z) \quad \text{(from (iii)}_n\text{)}$$
$$= \{ d(\theta x) \cdot a + (-1)^{|\theta x|} \theta x \cdot d(a) \} \cdot b \otimes z$$
$$+ (-1)^{|\theta x|+|a|} (\theta x \cdot a) d(b \otimes z) \quad \text{(from (ii))}.$$

On the other hand, it follows from (i) that

$$d(\theta x) \cdot (ab \otimes z) + (-1)^{|\theta x|} \theta x d(ab \otimes z)$$
$$= (d(\theta x) \cdot a) \cdot b \otimes z + (-1)^{|\theta x|} \theta x \{ d(a) \cdot b \otimes z + (-1)^{|a|} a \cdot d(b \otimes z) \}.$$

Therefore we see that if (i), (iii)$_n$ and (ii)$_n$ hold, then so does (ii)$_{n+1}$, and hence the equality (iii)$_{n+1}$ holds. By induction on n, we can deduce that the equality

(iii) $d(c \otimes \theta x \cdot a \otimes z) = d(c \otimes \theta x) \cdot a \otimes z + (-1)^{|c|+|\theta x|} c \otimes \theta x \cdot d(a \otimes z)$

holds for $a, c \in A$ and $z \in \overline{X}$.

Let (iv)$_n$ and (v)$_n$ denote the following equalities

$$d(\theta x_1 \cdots \theta x_n \cdot a \otimes z) = d(\theta x_1 \cdots \theta x_n) \cdot a \otimes z + (-1)^{|\theta x_1|+\cdots+|\theta x_n|} \theta x_1 \cdots \theta x_n \cdot d(a \otimes z)$$

for $a \in A$, $z \in \overline{X}$, and

$$d(c \otimes \theta x_1 \cdots \theta x_n \cdot a \otimes z) = d(c \otimes \theta x_1 \cdots \theta x_n) \cdot a \otimes z$$
$$+ (-1)^{|c|+|\theta x_1|+\cdots+|\theta x_n|} c \otimes \theta x_1 \cdots \theta x_n \cdot d(a \otimes z)$$

for any $a, c \in A$ and $z \in \overline{X}$ respectively. Suppose that the equality (iv)$_n$ holds. Then we have

$$d(c \otimes \theta x_1 \cdots \theta x_n \cdot a \otimes z)$$
$$= d(c \cdot (\theta x_1 \cdots \theta x_n \cdot a \otimes z))$$
$$= d(c) \cdot \theta x_1 \cdots \theta x_n \cdot a \otimes z + (-1)^{|c|} c \cdot d(\theta x_1 \cdots \theta x_n \cdot a \otimes z) \quad \text{(from (i))}$$
$$= d(c) \cdot \theta x_1 \cdots \theta x_n \cdot a \otimes z + (-1)^{|c|} c \cdot \{ d(\theta x_1 \cdots \theta x_n) \cdot a \otimes z$$

$$+ (-1)^{|\theta x_1 \cdots \theta x_n|} \theta x_1 \cdots \theta x_n \cdot d(a \otimes z)\} \quad \text{(from (iv)}_n\text{)}.$$

On the other hand

$$d(c \otimes \theta x_1 \cdots \theta x_n) \cdot a \otimes z + (-1)^{|c|+|\theta x_1|+\cdots+|\theta x_n|} c \otimes \theta x_1 \cdots \theta x_n \cdot d(a \otimes z)$$
$$= \{d(c) \cdot \theta x_1 \cdots \theta x_n + (-1)^{|c|} c \cdot d(\theta x_1 \cdots \theta x_n)\} \cdot (a \otimes z)$$
$$+ (-1)^{|c|+|\theta x_1 \cdots \theta x_n|} c \cdot ((\theta x_1 \cdots \theta x_n) \cdot d(a \otimes z)).$$

Thus we see that $(iv)_n$ implies $(v)_n$. Suppose that the equality $(iv)_{n-1}$ holds. Then it follows that

$$d(\theta x_1 \cdots \theta x_n \cdot a \otimes z) = d(\theta x_1 (\theta x_2 \cdots \theta x_n \cdot a \otimes z))$$
$$= d(\theta x_1) \cdot (\theta x_2 \cdots \theta x_n \cdot a \otimes z)$$
$$+ (-1)^{|\theta x_1|} \theta x_1 \cdot d(\theta x_2 \cdots \theta x_n \cdot a \otimes z) \quad \text{(from (iii))}$$
$$= d(\theta x_1) \cdot (\theta x_1 \cdots \theta_n \cdot a \otimes z)$$
$$+ (-1)^{|\theta x_1|} \theta x_1 \cdot \{d(\theta x_2 \cdots \theta x_n) \cdot a \otimes z$$
$$+ (-1)^{|\theta x_2 \cdots \theta x_n|} \theta x_2 \cdots \theta x_n \cdot d(a \otimes z)\} \quad \text{(from (iv)}_{n-1}\text{)}.$$

On the other hand,

$$d(\theta x_1 \cdots \theta x_n) \cdot a \otimes z + (-1)^{|\theta x_1 \cdots \theta x_n|} \theta x_1 \cdots \theta x_n \cdot d(a \otimes z)$$
$$= \{d(\theta x_1) \cdot \theta x_2 \cdots \theta x_n + (-1)^{|\theta x_1|} \theta x_1 d(\theta x_2 \cdots \theta x_n)\} \cdot a \otimes z$$
$$+ (-1)^{|\theta x_1 \cdots \theta x_n|} \theta x_1 (\theta x_2 \cdots \theta x_n \cdot d(a \otimes z)).$$

The equality $(iv)_{n-1}$ implies $(iv)_n$. By induction on n, we can conclude that the equality $(v)_n$ holds for any n. This completes the proof. □

PROOF OF LEMMA 5.12. We show that the equality (i) holds. Suppose that the following equality holds for any $a, b \in A$:

(•) $$d(a \cdot b) = d(a) \cdot b + (-1)^{|a|} a \cdot d(b).$$

Then we see

$$d(a \cdot b \otimes z) = d((a \cdot b) \otimes z)$$
$$= d(a \cdot b) \otimes z + (-1)^{|ab|} a \cdot b \otimes d(z)$$
$$= d(a) \cdot b \otimes z + (-1)^{|a|} a \cdot d(b) z + (-1)^{|ab|} a \cdot b \otimes d(z)$$
$$= d(a) \cdot b \otimes z + (-1)^{|a|} a \cdot d(b \otimes z).$$

5. A MULTIPLICATION ON A TWISTED TENSOR PRODUCT

We will verify that the equality ($\bullet$) holds. We write $\phi(a) = \sum_i a'_i \otimes a''_i$ and $\phi(b) = \sum_j b'_j \otimes b''_j$. Observe that ϕ is not the reduced diagonal map. Since

$$\phi(ab) = \sum_{i,j} (-1)^{|a''_i||b'_j|} a'_i b'_j \otimes a''_i b''_j,$$

it follows that

$$d(a \cdot b) = \sum_{i,j} (-1)^{|a''_i||b'_i| + |a'_i b'_j|} a'_i b'_j \otimes \theta(a''_i b''_j)$$

$$= \sum_{i,j, |a''_i| \geq 1} (-1)^{|a''_i||b'_i| + |a'_i b'_j|} a'_i b'_j \otimes \theta(a''_i b''_j) + \sum_j (-1)^{|a|+|b'_j|} ab'_j \otimes \theta(b''_j).$$

On the other hand, we have

$$d(a) \cdot b + (-1)^{|a|} a \cdot d(b)$$

$$= (\sum_i (-1)^{|a'_i|} a'_i \otimes \theta(a''_i)) \cdot b + (-1)^{|a|} a \cdot \sum_j (-1)^{|b'_j|} b'_j \otimes \theta b''_j$$

$$= \sum_{i, |a''_i| \geq 1} (-1)^{|a'_i|} a'_i \cdot (\theta(a''_i) \cdot b) + \sum_j (-1)^{|a|+|b'_j|} ab'_j \otimes \theta(b''_j)$$

$$= \sum_{i,j, |a''_i| \geq 1} (-1)^{|a'_i| + |\theta(a''_i)||b'_j|} a'_i b'_j \otimes \theta(a''_i b''_j) + \sum_j (-1)^{|a|+|b'_j|} ab'_j \otimes \theta(b''_j).$$

Thus we obtain the equality ($\bullet$).

By direct calculation, one can show that the equality (ii) holds. We write $\bar{\phi}(a'_i) = \sum_j b'_{ij} \otimes b''_{ij}$, $\bar{\phi}(x) = \sum_l x'_l \otimes x''_l$ and $\bar{\phi}(a''_i) = \sum_j \bar{b}_{ij} \otimes \bar{\bar{b}}_{ij}$. Then we have

$$d_W(\theta x \cdot a \otimes z)$$

$$= d_W(\sum_i (-1)^{|\theta x||a'_i|} a'_i \otimes \theta(xa''_i)z + (-1)^{|\theta x||a|} a \otimes \theta(x)z + 1 \otimes \theta(xa)z)$$

$$= \sum_i (-1)^{|\theta x||a'_i|} (\sum_j (-1)^{|b'_{ij}|} b'_{ij} \otimes \theta(b''_{ij})\theta(xa''_i)z + 1 \otimes \theta(a'_i)\theta(xa''_i)z)$$

$$+ \sum_i (-1)^{|\theta x||a'_i| + |a'_i|} a'_i \otimes \{d\theta(xa''_i)z + (-1)^{|\theta(xa''_i)|} \theta(xa''_i)dz\}$$

$$+ (-1)^{|\theta x||a|} \{\sum_i (-1)^{|a'_i|} a'_i \otimes \theta(a''_i)\theta(x)z + 1 \otimes \theta(a)\theta(x)z\}$$

$$+ (-1)^{|\theta x||a| + |a|} a \otimes \{d\theta(x)z + (-1)^{|\theta x|} \theta x dz\}$$

$$+ 1 \otimes d\theta(xa)z + (-1)^{|\theta(xa)|} 1 \otimes \theta(xa)dz.$$

Moreover we have

$$d\theta(xa) = -\{(-1)^{|x|} \theta x \theta a + \sum_i (-1)^{|x|+|a'_i|} \theta(xa'_i)\theta a''_i$$

$$+ (-1)^{|a||x|+|a|} \theta a \theta x \sum_i (-1)^{|x||a'_i|+|a'_i|} \theta a_i \theta(xa''_i)$$

$$+ \sum_l \sum_i (-1)^{|x_l''||a_i'|+|x_l'a_i'|}\theta(x_l'a_i')\theta(x_l''a_i'')$$
$$+ \sum_l (-1)^{|x_l''||a|+|x_l'a|}\theta(x_l'a)\theta(x_l'') + \sum_l (-1)^{|x_l'|}\theta x_l'\theta(x_l''a)\},$$

$$d\theta(xa_i'') = -\{(-1)^{|x|}\theta x \theta a_i'' + \sum_j (-1)^{|x|+|\bar{b}_{ij}|}\theta(x\bar{b}_{ij})\theta\bar{\bar{b}}_{ij}$$
$$+ (-1)^{|a_i''||x|+|a_i''|}\theta a_i''\theta x \sum_j (-1)^{|x||\bar{b}_{ij}|+|\bar{b}_{ij}|}\theta\bar{b}_{ij}\theta(x\bar{\bar{b}}_{ij})$$
$$+ \sum_l \sum_j (-1)^{|x_l''||\bar{b}_{ij}|+|x_l'\bar{b}_{ij}|}\theta(x_l'\bar{b}_{ij})\theta(x_l''\bar{\bar{b}}_{ij})$$
$$+ \sum_l (-1)^{|x_l''||a_i''|+|x_l'a_i''|}\theta(x_l'a_i'')\theta(x_l'') + \sum_l (-1)^{|x_l'|}\theta x_l'\theta(x_l''a_i'')\}.$$

On the other hand, we have

$$d(\theta x)\cdot a \otimes z + (-1)^{|\theta x|}\theta x \cdot d(a\otimes z)$$
$$= -\Big(\sum_l (-1)^{|x_l'|}\theta x_l'\theta x_l''\Big)\cdot a\otimes z$$
$$+ (-1)^{|\theta x|}\theta x\cdot\Big(\sum_i (-1)^{|a_i'|}a_i'\otimes\theta(a_i'')z + 1\otimes\theta(a)z + (-1)^{|a|}a\otimes dz\Big)$$
$$= -\sum_l (-1)^{|x_l'|}\theta x_l'\cdot\Big(\sum_i (-1)^{|\theta x_l''||a_i'|}a_i'\otimes\theta(x_l''a_i'')z$$
$$+ (-1)^{|\theta x_l''||a|}a\otimes\theta(x_l'')z + 1\otimes\theta(x_l''a)z\Big)$$
$$+ (-1)^{|\theta x|}\sum_i (-1)^{|a_i'|}\Big(\sum_j (-1)^{|\theta x||b_{ij}'|}b_{ij}'\otimes\theta(xb_{ij}'')\theta(a_i'')z$$
$$+ (-1)^{|\theta x||a_i'|}a_i'\otimes\theta(x)\theta(a_i'')z + 1\otimes\theta(xa_i')\theta a_i''z\Big)$$
$$+ (-1)^{|\theta x|}1\otimes\theta(x)\theta(a)z$$
$$+ (-1)^{|\theta x|+|a|}\Big(\sum_i (-1)^{|\theta x||a_i'|}a_i'\otimes\theta(xa_i'')dz$$
$$+ (-1)^{|\theta x||a|}a\otimes\theta x dz + 1\otimes\theta(xa)dz\Big)$$
$$= -\sum_l \sum_{i,j}(-1)^{|x_l'|+|\theta x_l''||a_i'|+|\theta x_l'||b_{ij}'|}b_{ij}'\otimes\theta(x_l'b_{ij}'')\theta(x_l''a_i'')z$$
$$- \sum_l \sum_i (-1)^{|x_l'|+|\theta x_l''||a_i'|+|\theta x_l'||a_i'|}a_i'\otimes\theta x_l'\theta(x_l''a_i'')z$$
$$- \sum_l \sum_i (-1)^{|x_l'|+|\theta x_l''||a_i'|}1\otimes\theta(x_l'a_i')\theta(x_l''a_i'')z$$
$$- \sum_l \sum_i (-1)^{|x_l'|+|\theta x_l''||a|+|\theta x_l'||a_i'|}a_i'\otimes\theta(x_l'a_i'')\theta(x_l'')z$$

5. A MULTIPLICATION ON A TWISTED TENSOR PRODUCT

$$-\sum_l (-1)^{|x'_l|+|\theta x''_l||a|+|\theta x'_l||a|} a \otimes \theta(x'_l)\theta(x''_l)z$$

$$-\sum_l (-1)^{|x'_l|+|\theta x''_l||a|} 1 \otimes \theta(x'_l a)\theta(x''_l)z$$

$$-\sum_l (-1)^{|x'_l|} 1 \otimes \theta(x'_l)\theta(x''_l a)z$$

$$+\sum_{i,j} (-1)^{|\theta x|+|a'_i|+|\theta x||b'_{ij}|} b'_{ij} \otimes \theta(xb''_{ij})\theta(a''_i)z$$

$$+\sum_i (-1)^{|\theta x|+|a'_i|+|\theta x||a'_i|} a'_i \otimes \theta(x)\theta(a''_i) \otimes z$$

$$+\sum_i (-1)^{|\theta x|+|a'_i|} 1 \otimes \theta(xa_i)\theta(a''_i) \otimes z$$

$$+(-1)^{|\theta x|} 1 \otimes \theta(x)\theta(a)z + \sum_i (-1)^{|\theta x|+|a|+|\theta x||a'_i|} a'_i \otimes \theta(xa''_i)dz$$

$$+(-1)^{|\theta x|+|a|+|\theta x||a|} a \otimes \theta(x)dz + (-1)^{|\theta x|+|a|} 1 \otimes \theta(xa)dz.$$

The same argument using the coassociativity of ϕ as in Proposition 5.6 works well in this case. Thus, by comparing the sign of each term, we have the equality (ii). □

When $A = L$, it follows from Ker $\bar{\theta} = 0$ that the multiplication m_W is well-defined. In this case, we see that the twisted tensor product (W, d_W, m_W) is nothing but the cobar resolution $D_A^*(\mathbb{Z}/p)$ of the left A-comodule $\mathbb{Z}/p$ endowed with an explicit pairing, which is defined as follows:

$$D_A^*(\mathbb{Z}/p) = \sum_{s=0}^\infty A \otimes \bar{A}^{\otimes s},$$

and the boundary operator $d_s : D_A^s(\mathbb{Z}/p) \to D_A^{s+1}(\mathbb{Z}/p)$ is given by

$$d_s(a_0 \otimes a_1 \otimes \cdots \otimes a_s) = (-1)^{|\alpha'_{0l}|} \alpha'_{0l} \otimes \alpha''_{0l}$$
$$+ \sum_{i=1}^s \sum_l (-1)^{\eta(i)} a_0 \otimes a_1 \otimes \cdots \otimes a_{i-1} \otimes \alpha'_{il} \otimes \alpha''_{il} \otimes a_{i+1} \cdots \otimes a_s,$$

where $\phi(a_i) = \sum_i a'_{il} \otimes a''_{il}$, $\eta(i) = \sum_{j=0}^{i-1} |a_i| + i + |\alpha'_{il}|$, $\bar{A} =$ Ker ε, $a_0 \in A$ and $a_1, \ldots, a_s \in \bar{A}$. The explicit associative pairing • is defined to be

$$(a_0 \otimes a_1 \otimes \cdots \otimes a_s) \bullet (b_0 \otimes b_1 \otimes \cdots \otimes b_t) = (-1)^k a_0 b_0^{(0)} \otimes a_1 b_0^{(1)} \otimes \cdots a_s b_0^{(s)} \otimes b_1 \otimes \cdots \otimes b_t,$$

where $k = \sum_{i=0}^s \deg b_0^{(i)} \left(s - i + \sum_{j=i+1}^s \deg a_j \right)$ and $b_0^{(0)} \otimes \cdots \otimes b_0^{(s)} \in A \otimes \bar{A}^{\otimes s}$ denotes the iterated coproduct of b_0.

REMARK 5.13. If $(A \otimes_\theta \bar{X}, d_W, m_W)$ is a well-defined differential graded algebra, then there exists an epimorphism of differential graded algebras $\pi : D_A^*(\mathbb{Z}/p) \to$

$A \otimes_\theta \overline{X}$ defined by
$$\pi(a_0 \otimes a_1 \otimes \cdots \otimes a_s) = a_0 \theta a_1 \cdots \theta a_s.$$

Moreover, when $A = H^*(G; \mathbb{Z}/p)$, we can obtain a morphism of graded algebras
$$\widetilde{\pi} = H(1 \square_A \pi) : \mathrm{Cotor}_A(A, \mathbb{Z}/p) \longrightarrow H(A \square (A \otimes_\theta \overline{X}), 1 \square d_W) = H(A \otimes \overline{X}, d).$$

Thus if $(A \otimes_\theta \overline{X}, d_W)$ is acyclic, then the morphism $\widetilde{\pi}$ is an isomorphism of algebras. The isomorphism plays a crucial role in determining the Steenrod operations in the E_2-term of the cobar EMSS converging to $H^*(BLSpin(10); \mathbb{Z}/2)$. For more details, see Appendix.

REMARK 5.14. When $A = L$, the differential graded algebra structure of the twisted tensor product coincides with that of the cobar resolution described in [**40**, A 1.2] *up to sign* of each term of the differential $d : D_A^*(\mathbb{Z}/p) \to D_A^{*+1}(\mathbb{Z}/p)$.

PROOF OF THEOREM 5.2. It follows from Lemma 5.10 that the multiplication m_W is associative. Therefore, to prove Theorem 5.2, it suffices to verify that
$$m_W(\psi \circ (\theta \otimes \theta) \circ \phi(ab) \otimes u) \equiv (-1)^{|\theta a||u|+|\theta b||u|} u \otimes \psi \circ (\theta \otimes \theta) \circ \phi(ab) \mod I$$

for any distinct elements a, b in $\{x_i\}$ and any element u in Q. We write $\bar{\phi}(a) = \sum_i a_i' \otimes a_i''$ and $\bar{\phi}(b) = \sum_j b_j' \otimes b_j''$.

LEMMA 5.15. *If $\theta(a_i'' u) \neq 0$ for some $u \in Q$, then $a \in Q^2$. Moreover, if $\theta(a_i' w) \neq 0$ for some $w \in \bar{A}$, then there exists an element $x^2 \in Q^2$ such that $u = x = w$ and $a = x^2$ up to constant.*

PROOF. If $a \in Q$, then we see from (III)(i) that $\theta(a_i'' u) = 0$. Therefore $a = x^2$ for some $x^2 \in Q^2$. Moreover, suppose that $\theta(a_i' w) \neq 0$. Since $\sum_i a_i' \otimes a_i'' = 2x \otimes x$, it follows from (IV) that $u = x = w$. □

LEMMA 5.16. *In the case $p = 2$ or 3, we have*
$$m_W(\psi \circ (\theta \otimes \theta) \circ \phi(x^3) \otimes x) = 0$$

for any $x^2 \in Q^2$.

PROOF. Since $\psi(x^3) = x^3 \otimes 1 + 3x^2 \otimes x + 3x \otimes x^2 + 1 \otimes x^3$, it follows that
$$m_W(\psi \circ (\theta \otimes \theta) \circ \phi(x^3) \otimes x) = 3(\theta(x^2) \otimes \theta x + \theta x \otimes \theta(x^2)) \cdot x$$
$$= 3(\theta(x^2) \otimes \theta(x^2) + \theta(x^2) \otimes \theta(x^2)) = 0. \quad \square$$

LEMMA 5.17. *For any a, b and $c \in \bar{A}$, we have $\theta(abc) = 0$.*

5. A MULTIPLICATION ON A TWISTED TENSOR PRODUCT

PROOF. Put $\{y_i\} = \{x_i\} \cap Q$. Then we can write

$$abc = \sum_{j_1,\ldots,j_l, l \geq 3} \alpha_{j_1,\ldots,j_l} y_{j_1} \cdots y_{j_l}$$

for some $\alpha_{j_1,\ldots,j_l} \in \mathbb{Z}/p$. From the definition of $\theta : A \to sL$ and (IV), it follows that

$$\theta(abc) = \sum_{j_1,\ldots,j_l, l \geq 3} \alpha_{j_1,\ldots,j_l} \theta(y_{j_1} \cdots y_{j_l}) = 0. \qquad \square$$

We continue the proof of Theorem 5.2. By the definition of the operator θ with degree $+1$, we see

$$D := m_W(\psi \circ (\theta \otimes \theta) \circ \phi(ab) \otimes u) - (-1)^{|\theta a||u|+|\theta b||u|} u \otimes \psi \circ (\theta \otimes \theta) \circ \phi(ab)$$

$$= \sum_{i,j} (-1)^{|a_i''||b_j'|+|a_i'b_j'|} \theta(a_i'b_j') \theta(a_i''b_j'') \cdot u + \sum_i (-1)^{|a_i''||b|+|a_i'b|} \theta(a_i'b) \theta(a_i'') \cdot u$$

$$+ \sum_i (-1)^{|a_i'|} \theta(a_i') \theta(a_i''b) \cdot u + \sum_j (-1)^{|ab_j'|} \theta(ab_j') \theta(b_j'') \cdot u$$

$$+ (-1)^{|a|} \theta(a) \theta(b) \cdot u + \sum_j (-1)^{|a||b_j'|+|b_j'|} \theta(b_j') \theta(ab_j'') \cdot u$$

$$+ (-1)^{|a||b|+|b|} \theta(b) \theta(a) \cdot u - (-1)^{|\theta a||u|+|\theta b||u|} u \otimes \psi \circ (\theta \otimes \theta) \circ \phi(ab).$$

By virtue of Lemma 5.16, we can assume that $ab \neq x^3$ or $u \neq x$ for any $x^2 \in Q^2$. Therefore, from Lemmas 5.15 and 5.17 and (IV), we have

$$D = (-1)^{|a|} (-1)^{|\theta b||u|} 1 \otimes \theta(au) \theta(b) + (-1)^{|a|} 1 \otimes \theta(a) \theta(bu)$$

$$+ (-1)^{|a||b|+|b|+|\theta a||u|} 1 \otimes \theta(bu) \theta a + (-1)^{|a||b|+|b|} 1 \otimes \theta b \theta(au).$$

Hence it follows from (IV) that $a = u \neq b$ or $b = u \neq a$ if $D \neq 0$ in X. In the case $a = u \neq b$, we see that

$$D = 1 \otimes \theta(a^2) \theta b + (-1)^{|b|} 1 \otimes \theta(b) \theta(a^2).$$

We can write $\phi(a^2) = a^2 \otimes 1 + 2a \otimes a + 1 \otimes a^2$ and $\phi(b) = b \otimes 1 + 1 \otimes b + \sum_i b_i' \otimes b_i''$. From (III)(i) and Lemma 5.17, it follows that $\theta(ab_i'') = \theta(a^2 b_i'') = \theta(a^2 b_i') = 0$ if $b \in Q$ or $b \neq a^2$. Therefore the condition (IV) allows us to deduce that $\theta(ab) = 0$. Thus we see that $\theta(a^2) \theta b + (-1)^{|b|} \theta(b) \theta(a^2) \in I$. The same argument also works in the case $b = u \neq a$. It turns out that $\psi \circ (\theta \otimes \theta) \circ \phi(ab) \cdot u = 0$ in W. $\qquad \square$

PROOF OF THEOREM 5.3. We recall from [34] the Hopf algebra structure of $H^*(E_8; \mathbb{Z}/5)$:

$$H^*(E_8; \mathbb{Z}/5) \cong \mathbb{Z}/5[x_{12}]/(x_{12}^5) \otimes \Lambda(x_3, x_{11}, x_{15}, x_{23}, x_{27}, x_{35}, x_{39}, x_{47})$$

with x_i primitive for $i = 3, 11, 12$, and

$$\bar{\phi}(x_j) = x_{12} \otimes x_{j-12} \qquad \text{for } j = 15, 23,$$

$$\bar{\phi}(x_k) = 2x_{12} \otimes x_{k-12} + x_{12}^2 \otimes x_{k-24} \qquad \text{for } k = 27, 35,$$

$$\bar{\phi}(x_l) = 3x_{12} \otimes x_{l-12} + 3x_{12}^2 \otimes x_{l-24} + x_{12}^3 \otimes x_{l-36} \qquad \text{for } l = 39, 47.$$

Let L denote the $\mathbb{Z}/5$-subspace of $A = H^*(E_8; \mathbb{Z}/5)$ which gives rise to the twisted tensor product $A \otimes_\theta \overline{X}$ in [34], that is, L is the subspace generated by

$$\tilde{L} := \{x_{12}, x_{12}^2, x_{12}^3, x_{12}^4, x_3, x_{11}, x_{15}, x_{23}, x_{27}, x_{35}, x_{39}, x_{47}\}.$$

In order to show that the multiplication m_W is well-defined, it suffices to verify that

$$D := m_W(\psi \circ (\theta \otimes \theta) \circ \phi(ab) \otimes u) - (-1)^{|\theta a||u|+|\theta b||u|} u \otimes \psi \circ (\theta \otimes \theta) \circ \phi(ab)$$

is an element in I for any $u \in Q$, $a \in M := \{x_{12}, x_{12}^2, x_{12}^3, x_{12}^4\}$ and $b \in \tilde{L}\setminus M$ or $a, b \in \tilde{L}\setminus M$. We write $\bar{\phi}(a) = \sum_i a_i' \otimes a_i''$ and $\bar{\phi}(b) = \sum_j b_j' \otimes b_j''$. Since $\theta(a_i'b) = \theta(a_i''b_j'') = \theta(a_i''b) = \theta(ab_j'') = 0$ and $\theta(b_j''u) = 0$, it follows from the formula of D in the proof of Theorem 5.2 that

$$D = (-1)^{|a|}\theta(a)\theta(b) \cdot u + (-1)^{|a||b|+|b|}\theta b \theta a \cdot u + \sum_j (-1)^{|ab_j'|}\theta(ab_j')\theta(b_j'') \cdot u$$

$$- (-1)^{|\theta a||u|+|\theta b||u|} u \otimes \Big((-1)^{|a|}\theta(a)\theta(b) + (-1)^{|a||b|+|b|}\theta b \theta a$$

$$+ \sum_j (-1)^{|ab_j'|}\theta(ab_j')\theta(b_j'') \Big)$$

$$= (-1)^{|a|}\theta a \cdot \big((-1)^{|\theta b||u|} u \theta(b) \big) + (-1)^{|a||b|+|b|}\theta b \cdot \big((-1)^{|\theta a||u|} u \theta a + 1 \otimes \theta(au) \big)$$

$$+ \sum_j (-1)^{|ab_j'|+|\theta b_j''||u|} \theta(ab_j') \cdot u\theta(b_j'')$$

$$- (-1)^{|\theta a||u|+|\theta b||u|+|a|} u \theta a \theta b - (-1)^{|\theta a||u|+|\theta b||u|+|a||b|+|b|} u \theta b \theta a$$

$$- (-1)^{|\theta a||u|+|\theta b||u|} \sum_j (-1)^{|ab_j'|} u \theta(ab_j')\theta(b_j'')$$

$$= (-1)^{|a|+|\theta b||u|} \big((-1)^{|\theta a||u|} u \theta a + \theta(au) \big) \theta b$$

$$+ (-1)^{|a||b|+|b|+|\theta a||u|+|\theta b||u|} u \theta b \theta a + (-1)^{|a||b|+|b|}\theta b \theta(au)$$

$$+ \sum_j (-1)^{|ab_j'|+|\theta b_j''||u|} \{ (-1)^{|\theta(ab_j')||u|} u\theta(ab_j') + \theta(ab_j'u) \} \theta b_j''$$

$$- (-1)^{|\theta a||u|+|\theta b||u|+|a|} u \theta a \theta b - (-1)^{|\theta a||u|+|\theta b||u|+|a||b|+|b|} u \theta b \theta a$$

$$- \sum_j (-1)^{|\theta a||u|+|\theta b||u|+|ab_j'|} u \theta(ab_j')\theta(b_j'')$$

$$= 1 \otimes \theta(au)\theta b + (-1)^{|a||b|+|b|}\theta b \theta(au) + \sum_j (-1)^{|ab_j'|+|\theta b_j''||u|} \theta(ab_j'u)\theta b_j''$$

$$= \psi \circ (\theta \otimes \theta) \circ \Delta((au)b) \in I.$$

Thus we see that m_W is well-defined in $A \otimes_\theta \overline{X}$. □

PROOF OF THEOREM 5.4. Let us recall from [**24**] the Hopf algebra structure of $H^*(E_8; \mathbb{Z}/2)$:

$$H^*(E_8; \mathbb{Z}/2) \cong \mathbb{Z}/2[x_3, x_5, x_9, x_{15}]/(x_3^{16}, x_5^8, x_9^4, x_{15}^4) \otimes \Lambda(x_{17}, x_{23}, x_{27}, x_{29})$$

with x_i primitive for $i = 3, 5, 9, 17$, and

$$\bar{\phi}(x_{15}) = x_3 \otimes x_3^4 + x_5 \otimes x_5^2 + x_3^2 \otimes x_9,$$
$$\bar{\phi}(x_{23}) = x_3 \otimes x_3^4 + x_5 \otimes x_9^2 + x_3^2 \otimes x_{17},$$
$$\bar{\phi}(x_{27}) = x_3 \otimes x_3^8 + x_9 \otimes x_9^2 + x_5^2 \otimes x_{17},$$
$$\bar{\phi}(x_{29}) = x_5 \otimes x_3^8 + x_9 \otimes x_5^4 + x_3^4 \otimes x_{17}.$$

Put $M = \{x_3, x_5, x_9, x_{15}, x_{17}, x_{23}, x_{27}, x_{29}\}$. Let u be an element in M and α an element on A with $\theta(\alpha) = 0$. We write $\bar{\phi}(\alpha) = \sum_i \alpha'_i \otimes \alpha''_i$ and $\bar{\phi}(u) = \sum_s u'_s \otimes u''_s$. Since u'_s is primitive, it follows that

$$D := m_W(\psi \circ (\theta \otimes \theta) \circ \phi(\alpha) \otimes u) - u \otimes \psi \circ (\theta \otimes \theta) \circ \phi(\alpha)$$
$$= \sum_i \theta\alpha'_i \theta\alpha''_i \cdot u + u \otimes \sum_i \theta\alpha'_i \theta\alpha''_i$$
$$= \sum_{i,s} \theta\alpha'_i \cdot u'_s \theta(\alpha''_i u''_s) + \sum_i \theta\alpha'_i \cdot u\theta(\alpha''_i) + \sum_i \theta\alpha'_i \theta(\alpha''_i u) + u \sum_i \theta\alpha'_i \theta\alpha''_i$$
$$= \sum_{i,s} u'_s \theta\alpha'_i \theta(\alpha''_i u''_s) + \sum_{i,s} \theta(\alpha'_i u'_s) \theta(\alpha''_i u''_s) + \sum_i \theta(\alpha'_i u) \theta\alpha''_i$$
$$+ \sum_i u\theta(\alpha'_i)\theta(\alpha''_i) + \sum_{i,s} u'_s \theta(\alpha'_i u''_s) \theta(\alpha''_i) + \sum_i \theta(\alpha'_i) \theta(\alpha''_i u) + u \sum_i \theta\alpha'_i \theta\alpha''_i.$$

Since the element α belongs to Ker θ, so does αx for any $x \in A$. Therefore, in W, we have

$$0 = \psi \circ (\theta \otimes \theta) \circ \phi(\alpha u) = \sum_{i,s} \theta(\alpha'_i u'_s) \theta(\alpha''_i u''_s) + \sum_i \theta(\alpha'_i u) \theta(\alpha''_i) + \sum_i \theta(\alpha'_i) \theta(\alpha''_i u).$$

Since u''_s is also primitive, it follows that

$$0 = \psi \circ (\theta \otimes \theta) \circ \phi(\alpha u''_s) = \sum_i \theta(\alpha'_i u''_s) \theta(\alpha''_i) + \sum_i \theta(\alpha'_i) \theta(\alpha''_i u''_s)$$

in W for any u''_s. These facts imply that $D = 0$ in W, and hence m_W is well-defined. $\square$

The epimorphism π in the proof of Theorem 5.1 is also a morphism of $H^*(G; \mathbb{Z}/p)$-comodules. As an immediate consequence, we have Proposition 5.6.

This section is concluded with some examples of differential graded algebras $A \square_A (A \otimes_\theta \overline{X})$ for computing the algebras $\mathrm{Cotor}_A(A, \mathbb{Z}/p)$. We can see the original twisted tensor product in the cited paper:

The case $(E_6, 2)$, [**22**];

$$W' = A\square_A(A \otimes_\theta \overline{X}) = \mathbb{Z}/2[x_3]/(x_3^4) \otimes \Lambda(x_5, x_9, x_{15}, x_{17}, x_{23})$$
$$\otimes \mathbb{Z}/2\{\theta x_3, \theta(x_3^2), \theta x_5, \theta x_9, \theta x_{17}, \theta x_{15}, \theta x_{23}\}/I$$
$$= \mathbb{Z}/2[x_3]/(x_3^4) \otimes \Lambda(x_5, x_9, x_{15}, x_{17}, x_{23}) \otimes \mathbb{Z}/2[a_4, a_7, a_6, a_{10}, a_{18}, b_{16}, b_{24}],$$

$$d(x_j) = x_{j-6} \otimes a_7 + x_3^2 \otimes a_{j-6+1} \ (j = 15, 23),$$
$$d|_{\mathbb{Z}/2\{\ \}/I} = \text{ the ordinary differential on } \mathbb{Z}/2\{\ \}/I,$$

where $a_{i+1} = \theta x_i$ for $i = 3, 5, 7, 9,$ and 17, $a_7 = \theta(x_3^2)$, $b_{16} = \theta x_{15}$ and $b_{24} = \theta x_{23}$. Observe that, in this case, W' is a commutative differential graded algebra.

The case $(E_7, 2)$, [**24**];

$$W' = A\square_A(A \otimes_\theta \overline{X})$$
$$= \mathbb{Z}/2[x_3, x_5, x_9]/(x_3^4, x_5^4, x_9^4) \otimes \Lambda(x_{15}, x_{17}, x_{23}, x_{27})$$
$$\otimes \mathbb{Z}/2\{\theta x_3, \theta x_3^2, \theta x_5, \theta x_5^2, \theta x_9, \theta x_9^2, \theta x_{17}, \theta x_{15}, \theta x_{23}, \theta x_{27}\}/I,$$
$$d(x_{15}) = x_5^2 \otimes \theta x_5 + x_5 \otimes \theta(x_5^2) + x_3^2 \otimes \theta x_9 + x_9 \otimes \theta(x_3^2),$$
$$d(x_{23}) = x_9^2 \otimes \theta x_5 + x_5 \otimes \theta(x_9^2) + x_3^2 \otimes \theta x_{17} + x_{17} \otimes \theta(x_3^2),$$
$$d(x_{27}) = x_9^2 \otimes \theta x_9 + x_9 \otimes \theta(x_9^2) + x_5^2 \otimes \theta x_{17} + x_{17} \otimes \theta(x_5^2),$$
$$d|_{\mathbb{Z}/2\{\ \}/I} = \text{ the ordinary differential on } \mathbb{Z}/2\{\ \}/I,$$
$$\theta x_5 \cdot x_{15} = x_{15} \otimes \theta x_5 + x_5^2 \otimes \theta(x_5^2),$$
$$\theta x_9 \cdot x_{15} = x_{15} \otimes \theta x_9 + x_3^2 \otimes \theta(x_9^2),$$
$$\theta x_5 \cdot x_{23} = x_{23} \otimes \theta x_5 + x_9^2 \otimes \theta(x_5^2),$$
$$\theta x_9 \cdot x_{27} = x_{27} \otimes \theta x_9 + x_9^2 \otimes \theta(x_9^2).$$

The case $(E_8, 2)$, [**31**];

$$W' = A\square_A(A \otimes_\theta \overline{X})$$
$$= \mathbb{Z}/2[x_3, x_5, x_9, x_{15}]/(x_3^{16}, x_5^8, x_9^4, x_{15}^4) \otimes \Lambda(x_{17}, x_{23}, x_{27}, x_{29})$$
$$\otimes \mathbb{Z}/2\{\theta(x_3^i), \theta(x_5^j), \theta x_{15}, \theta x_{15}^2, \theta x_{17}, \theta x_{23}, \theta x_{27}, \theta x_{29}, \theta(x_3^k x_9) \ ;$$
$$1 \le i \le 8, 1 \le j \le 4, 1 \le k \le 7\}/I,$$

$$d(x_{15}) = x_3^4 \otimes \theta x_3 + x_3 \otimes \theta(x_3^4) + x_5^2 \otimes \theta x_5 + x_5 \otimes \theta(x_5^2) + x_3^2 \otimes \theta x_9 + x_9 \otimes \theta(x_3^2),$$
$$d(x_{23}) = x_5^4 \otimes \theta x_3 + x_3 \otimes \theta(x_5^4) + x_9^2 \otimes \theta x_5 + x_5 \otimes \theta(x_9^2) + x_3^2 \otimes \theta x_{17} + x_{17} \otimes \theta(x_3^2),$$
$$d(x_{27}) = x_3^8 \otimes \theta x_3 + x_3 \otimes \theta(x_3^8) + x_9^2 \otimes \theta x_9 + x_9 \otimes \theta(x_9^2) + x_5^2 \otimes \theta x_{17} + x_{17} \otimes \theta(x_5^2),$$
$$d(x_{29}) = x_3^8 \otimes \theta x_5 + x_5 \otimes \theta(x_3^8) + x_5^4 \otimes \theta x_9 + x_9 \otimes \theta(x_5^4) + x_3^4 \otimes \theta x_{17} + x_{17} \otimes \theta(x_3^4),$$

5. A MULTIPLICATION ON A TWISTED TENSOR PRODUCT

$d|_{\mathbb{Z}/2\{\ \}/I} = $ the ordinary differential on $\mathbb{Z}/2\{\ \}/I$,

$\theta x_3^i \cdot x_{15} = x_3^4 \otimes \theta(x_3^{i+1}) + x_3 \otimes \theta(x_3^{i+4}) + x_9 \otimes \theta(x_3^{i+2}) + x_3^2 \otimes \theta(x_3^i x_9)$
$\qquad + x_{15} \otimes \theta(x_3^i),$

$\theta x_5^i \cdot x_{15} = x_{15} \otimes \theta x_5^i + x_5^2 \otimes \theta(x_5^{i+1}) + x_5 \otimes \theta(x_5^{i+2}),$

$\theta x_9 \cdot x_{15} = x_{15} \otimes \theta x_9 + x_3^4 \otimes \theta(x_3 x_9) + x_3 \theta(x_3^4 x_9) + x_9 \otimes \theta(x_3^2 x_9) + x_3^2 \theta(x_9^2),$

$\theta x_3^i \cdot x_{23} = x_{23} \otimes \theta x_3^i + x_5^4 \otimes \theta(x_3^{i+1}) + x_{17} \otimes \theta x_3^{i+2},$

$\theta x_9 \cdot x_{23} = x_{23} \otimes \theta x_9 + x_5^4 \otimes \theta(x_3 x_9) + x_{17} \otimes \theta(x_3^2 x_9),$

$\theta x_5^i \cdot x_{23} = x_{23} \otimes \theta x_5^i + x_9^2 \otimes \theta(x_5^{i+1}),$

$\theta x_3^i \cdot x_{27} = x_{27} \otimes \theta x_3^i + x_3^8 \otimes \theta(x_3^{i+1}) + x_9^2 \otimes \theta(x_3 x_9),$

$\theta x_5^i \cdot x_{27} = x_{27} \otimes \theta x_5^i + x_{17} \otimes \theta(x_5^{i+2}),$

$\theta x_9 \cdot x_{27} = x_{27} \otimes \theta x_9 + x_3^8 \otimes \theta(x_3 x_9) + x_9^2 \otimes \theta(x_9^2),$

$\theta x_3^i \cdot x_{29} = x_{29} \otimes \theta x_3^i + x_5^4 \otimes \theta(x_3^i x_9) + x_{17} \otimes \theta x_3^{i+4},$

$\theta x_5^i \cdot x_{29} = x_{29} \otimes \theta x_5^i + x_3^8 \otimes \theta(x_5^{i+1}),$

$\theta x_9 \cdot x_{29} = x_{29} \otimes \theta x_9 + x_5^4 \otimes \theta(x_9^2) + x_{17} \otimes \theta(x_3^4 x_9),$

$\theta(x_3^i x_9) \cdot x_{15} = x_{15} \otimes \theta(x_3^i x_9) + x_3^4 \otimes \theta(x_3^{i+1} x_9) + x_3 \otimes \theta(x_3^{i+4} x_9) + x_9 \otimes \theta(x_3^{i+2} x_9),$

$\theta(x_3^i x_9) \cdot x_{23} = x_{23} \otimes \theta(x_3^i x_9) + x_{17} \otimes \theta(x_3^{i+2} x_9),$

$\theta(x_3^i x_9) \cdot x_{27} = x_{27} \otimes \theta(x_3^i x_9) + x_3^8 \otimes \theta(x_3^{i+1} x_9),$

$\theta(x_3^i x_9) \cdot x_{29} = x_{29} \otimes \theta(x_3^i x_9) + x_{17} \otimes \theta(x_3^{i+4} x_9).$

The case $(G,p) = (PU(3), 3)$, [**23**];

$W' = A\square_A(A \otimes_\theta \overline{X}) = \mathbb{Z}/3[x_2]/(x_2^3) \otimes \Lambda(x_1, x_3) \otimes \mathbb{Z}/3\{a_2, a_3, c_5, b_4\}/I,$

$$db_4 = -a_2 a_3, \qquad dc_5 = a_3^2,$$
$$d(x_3) = x_2 \otimes a_2 + x_1 \otimes a_3,$$
$$a_3 \cdot x_3 = -x_3 \otimes a_3 + x_1 \otimes c_5.$$

The case $(G,p) = (F_4, 3)$, [**23**];

$W' = A\square_A(A \otimes_\theta \overline{X})$
$\qquad = \mathbb{Z}/3[x_8]/(x_8^3) \otimes \Lambda(x_3, x_7, x_{11}, x_{15}) \otimes \mathbb{Z}/3\{a_4, a_8, a_9, b_{12}, b_{16}, c_{17}\}/I,$

$$d(x_j) = x_8 \otimes a_{j-8+1} + x_{j-8} \otimes a_9 \ (j = 11, 15),$$
$d|_{\mathbb{Z}/3\{\ \}/I} = $ the ordinary differential on $\mathbb{Z}/3\{\ \}/I$,

$$a_9 \cdot x_j = -x_j \otimes a_9 + x_{j-8} \otimes c_{17} \ (j = 11, 15).$$

The case $(G, p) = (E_6, 3)$, [**32**];

$$W' = A\square_A(A \otimes_\theta \overline{X}) = \mathbb{Z}/3[x_8]/(x_8^3) \otimes \Lambda(x_3, x_7, x_9, x_{11}, x_{15}, x_{17})$$
$$\otimes \mathbb{Z}/3\{a_4, a_8, a_9, a_{10}, b_{12}, b_{16}, b_{18}, c_{17}\}/I,$$

$$d(x_j) = x_8 \otimes a_{j-8+1} + x_{j-8} \otimes a_9 \ (j = 11, 15, 17),$$
$$d|_{\mathbb{Z}/3\{\ \}/I} = \text{ the ordinary differential on } \mathbb{Z}/3\{\ \}/I,$$
$$a_9 \cdot x_j = -x_j \otimes a_9 + x_{j-8} \otimes c_{17} \ (j = 11, 15, 17).$$

The case $(E_7, 3)$, [**32**];

$$W' = A\square_A(A \otimes_\theta \overline{X}) = \mathbb{Z}[x_8]/(x_8^3) \otimes \Lambda(x_3, x_7, x_{11}, x_{15}, x_{19}, x_{27}, x_{35})$$
$$\otimes \mathbb{Z}/3\{a_4, a_8, a_9, a_{20}, b_{12}, b_{16}, b_{28}, c_{17}, e_{36}\}/I,$$

$$d(x_j) = x_8 \otimes a_{j-8+1} + x_{j-8} \otimes a_9 \ (j = 11, 15, 27),$$
$$d(x_{35}) = x_8 \otimes b_{28} + x_{27} \otimes a_9 - x_8^2 \otimes a_{20} + x_{19} \otimes c_{17},$$
$$d|_{\mathbb{Z}/3\{\ \}/I} = \text{ the ordinary differential on } \mathbb{Z}/3\{\ \}/I,$$
$$a_9 \cdot x_j = -x_j \otimes a_9 + x_{j-8} \otimes c_{17} \ (j = 11, 15, 27, 35).$$

The case $(E_8, 3)$, [**33**];

$$W' = A\square_A(A \otimes_\theta \overline{X})$$
$$= \mathbb{Z}/3[x_8, x_{20}]/(x_8^3, x_{20}^3) \otimes \Lambda(x_3, x_7, x_{15}, x_{19}, x_{27}, x_{35}, x_{39}, x_{47})$$
$$\otimes \mathbb{Z}/3\{a_4, a_8, a_9, a_{20}, a_{21}, c_{17}, c_{41}, b_{16}, b_{40}, d_{28}, e_{36}, e_{48}\}/I,$$

$$d(x_{15}) = x_8 \otimes a_8 + x_7 \otimes a_9, \qquad d(x_{39}) = x_{20} \otimes a_{20} + x_{19} \otimes a_{21},$$
$$d(x_{27}) = x_8 \otimes a_{20} + x_{19} \otimes a_9 + x_{20} \otimes a_8 + x_7 \otimes a_{21},$$
$$d(x_{35}) = x_8 \otimes d_{28} + x_{27} \otimes a_9 - x_8^2 \otimes a_{20} - x_{19} \otimes c_{17} + x_{20} \otimes b_{16} + x_{15} \otimes a_{21}$$
$$+ x_{20}x_8 \otimes a_8,$$
$$d(x_{47}) = x_8 \otimes b_{40} + x_{39} \otimes a_8 + x_{20} \otimes d_{28} + x_{27} \otimes a_{21} + x_7 \otimes c_{41} - x_{20}^2 \otimes a_8$$
$$+ x_{20}x_8 \otimes a_{20},$$
$$d|_{\mathbb{Z}/3\{\ \}/I} = \text{ the ordinary differential on } \mathbb{Z}/3\{\ \}/I,$$

$$a_9 \cdot x_{15} = -x_{15} \otimes a_9 + x_7 \otimes c_{17}, \qquad a_{21} \cdot x_{39} = -x_{39} \otimes a_{21} + x_{19} \otimes c_{41},$$
$$a_9 \cdot x_{27} = -x_{27} \otimes a_9 + x_{19} \otimes c_{17}, \qquad a_{21} \cdot x_{27} = -x_{27} \otimes a_{21} + x_7 \otimes c_{41},$$

$$a_9 \cdot x_{35} = -x_{35} \otimes a_9 + x_{27} \otimes c_{17}, \qquad a_{21} \cdot x_{35} = -x_{35} \otimes a_{21} + x_{15} \otimes c_{41},$$
$$a_9 \cdot x_{47} = -x_{47} \otimes a_9 + x_{39} \otimes c_{17}, \qquad a_{21} \cdot x_{47} = -x_{47} \otimes a_{21} + x_{27} \otimes c_{41}.$$

The case $(E_8, 5)$, [**34**];

$$W' = A\square_A(A \otimes_\theta \overline{X})$$
$$= \mathbb{Z}/3[x_{12}]/(x_{12}^5) \otimes \Lambda(x_3, x_{11}, x_{15}, x_{23}, x_{27}, x_{35}, x_{39}, x_{47})$$
$$\otimes \mathbb{Z}/3\{\theta x_3, \theta x_{11}, \theta(x_{12}^i), \theta x_{15}, \theta x_{23}, \theta x_{27}, \theta x_{35}, \theta x_{39}, \theta x_{47},\ ;\ 1 \le i \le 4\}/I,$$

$$d(x_j) = x_{j-12} \otimes \theta x_{12} + x_{12} \otimes \theta_{j-12}, \qquad \text{for } j = 15, 13,$$

$$d(x_k) = -2x_{k-12} \otimes \theta x_{12} + 2x_{12} \otimes \theta x_{k-12}$$
$$+ x_{k-24} \otimes \theta(x_{12}^2) + x_{12}^2 \otimes \theta x_{k-24}, \qquad \text{for } k = 27, 35,$$

$$d(x_l) = 3x_{12} \otimes \theta x_{l-12} + 3x_{12}^2 \otimes \theta x_{l-24} + x_{12}^3 \otimes \theta x_{l-36}$$
$$- 3x_{l-12} \otimes \theta x_{12} + 3x_{l-24} \otimes \theta x_{12}^2 - x_{l-36} \otimes \theta(x_{12}^3), \qquad \text{for } l = 39, 47,$$

$$d|_{\mathbb{Z}/3\{\ \}/I} = \text{the ordinary differential on } \mathbb{Z}/3\{\ \}/I,$$

$$\theta x_{12} \cdot x_k = -x_k \otimes \theta x_{12} + 2x_{k-12} \otimes \theta(x_{12}^2) - x_{k-24} \otimes \theta(x_{12}^3), \qquad \text{for } k = 27, 35$$

$$\theta x_{12} \cdot x_l = -x_l \otimes \theta x_{12} + 3x_{l-12} \otimes \theta(x_{12}^2) - 3x_{k-24} \otimes \theta(x_{12}^3)$$
$$+ x_{l-36} \otimes \theta(x_{12}^4), \qquad \text{for } l = 39, 47.$$

The differential operator d and the bracket [,] are trivial on the generators if they are not indicated above.

6. The twisted tensor product associated with $H^*(Spin(N); \mathbb{Z}/2)$

Before Kono, Mimura and Shimada started the computation of the cohomology of classifying spaces of Lie groups by using twisted tensor products, Quillen [**39**] had determined explicitly the algebra structure of $H^*(BSpin(N); \mathbb{Z}/2)$. Due to this fact, a twisted tensor product associated with $H^*(Spin(N); \mathbb{Z}/2)$ has been unconstructed. We begin with a lemma to construct the twisted tensor product in this section.

LEMMA 6.1. *Let $A = \mathbb{Z}/2[x]/(x^{2^r})$ be the Hopf algebra with a primitive element x. Put $L = \mathbb{Z}/2\{x, x^2, \ldots, x^{2^{r-1}-1}, x^{2^{r-1}}\}$. Then the twisted tensor product $W = A \otimes_\theta T(sL)/I$ forms an injective resolution $0 \to \mathbb{Z}/2 \to W$ of $\mathbb{Z}/2$ as an A-comodule.*

PROOF. We first prove that the multiplication m_W on W stated in Theorem 5.1 is well-defined. Recall from Section 5 the definition of the ideal I:

$$I = (\psi \circ (\bar{\theta} \otimes \bar{\theta}) \circ \phi_A)(\text{Ker } \bar{\theta}),$$

where $\bar{\theta} : A \to sL$ is the natural projection, ϕ_A and ψ are the coproduct of A and the product of $T(sL)$, respectively. Observe that

$$\operatorname{Ker} \bar{\theta} \cong \mathbb{Z}/2\{x^{2^{r-1}+1}, x^{2^{r-1}+2}, \ldots, x^{2^r-1}\}$$

as a vector space. Since the element $x^{2^{r-1}}$ is primitive, it follows that, for $1 \leq l \leq 2^{r-1} - 1$,

$$\phi_A(x^{2^{r-1}} x^l) = (x^{2^{r-1}} \otimes 1 + 1 \otimes x^{2^{r-1}})(x^l \otimes 1 + 1 \otimes x^l + \sum_{i=1}^{l-1} \binom{l}{i} x^{l-i} \otimes x^i)$$

$$= x^{2^{r-1}} x^l \otimes 1 + x^{2^{r-1}} \otimes x^l + x^l \otimes x^{2^{r-1}} + 1 \otimes x^{2^{r-1}} x^l$$

$$+ \sum_{i=1}^{l-1} \binom{l}{i} x^{2^{r-1}+l-i} \otimes x^i + \sum_{i=1}^{l-1} \binom{l}{i} x^{l-i} \otimes x^{2^{r-1}+i},$$

and hence

$$(\psi \circ (\bar{\theta} \otimes \bar{\theta}) \circ \phi_A)(x^{2^{r-1}} x^l) = \theta x^{2^{r-1}} \theta x^l + \theta x^l \theta x^{2^{r-1}}.$$

Therefore we see that the ideal I is generated by $\theta x^{2^{r-1}} \theta x^l + \theta x^l \theta x^{2^{r-1}}$ for $1 \leq l \leq 2^{r-1} - 1$. This fact implies that W is isomorphic to $A \otimes T(sL') \otimes \mathbb{Z}/2[\theta x^{2^{r-1}}]$ as an algebra, where $L' = \mathbb{Z}/2\{x, x^2, \ldots, x^{2^{r-1}-1}\}$. The multiplication m_W is compatible with the associative pairing on the cobar resolution $D_A^*(\mathbb{Z}/2)$ under the projection $D_A^*(\mathbb{Z}/2) \to W$. Hence, in order to verify that the multiplication m_W is well-defined, it suffices to prove that $m_W(u \otimes x)$ is in $A \otimes I$ for any $u \in I$. Since $\theta x^i \cdot x = x\theta x^i + \theta x^{i+1}$ for $1 \leq i \leq 2^{r-1}$, it follows that, modulo I,

$$\theta x^{i_1} \cdots \theta x^{i_s} (\theta x^{2^{r-1}} \theta x^l + \theta x^l \theta x^{2^{r-1}}) \theta x^{j_1} \cdots \theta x^{j_q} \cdot x$$

$$\equiv \theta x^{i_1} \cdots \theta x^{i_s} (\theta x^{2^{r-1}} \theta x^l + \theta x^l \theta x^{2^{r-1}}) x \theta x^{j_1} \cdots x \theta x^{j_q}$$

$$\equiv \theta x^{i_1} \cdots \theta x^{i_s} (\theta x^{2^{r-1}} \cdot (x\theta x^l + \theta x^{l+1}) + \theta x^l \cdot x\theta x^{2^{r-1}}) \theta x^{j_1} \cdots \theta x^{j_q}$$

$$\equiv \theta x^{i_1} \cdots \theta x^{i_s} \cdot x(\theta x^{2^{r-1}} \theta x^l + \theta x^l \theta x^{2^{r-1}}) \theta x^{j_1} \cdots \theta x^{j_q}$$

$$+ \theta x^{i_1} \cdots \theta x^{i_s} (\theta x^{2^{r-1}} \theta x^{l+1} + \theta x^{l+1} \theta x^{2^{r-1}}) \theta x^{j_1} \cdots \theta x^{j_q}$$

$$\equiv x\theta x^{i_1} \cdots \theta x^{i_s} (\theta x^{2^{r-1}} \theta x^l + \theta x^l \theta x^{2^{r-1}}) \theta x^{j_1} \cdots \theta x^{j_q}$$

$$\equiv 0.$$

It turns out that m_W is well-defined. There is an isomorphism of coalgebras

$$A = \mathbb{Z}/2[x]/(x^{2^r}) \cong \mathbb{Z}/2[x]/(x^{2^{r-1}}) \otimes \Lambda(x^{2^{r-1}}) =: B \otimes C.$$

We now define a map

$$\eta : D_B^*(\mathbb{Z}/2) \otimes D_C^*(\mathbb{Z}/2) = B \otimes T(sL') \otimes C \otimes \mathbb{Z}/2[\theta x^{2^{r-1}}]$$

$$\longrightarrow A \otimes T(sL') \otimes \mathbb{Z}/2[\theta x^{2^{r-1}}] = W$$

by $\eta(\alpha \otimes \beta) = \alpha \cdot \beta$. Since the element $x^{2^{r-1}}$ is primitive, we see that

$$\theta x^i \cdot x^{2^{r-1}} = x^{2^{r-1}} \theta x^i + \theta x^{2^{r-1}+i} = x^{2^{r-1}} \theta x^i.$$

Hence we have

$$\eta((b \otimes X) \otimes (x^{2^{r-1}} \otimes Y)) = bx^{2^{r-1}} \otimes XY.$$

This fact implies that η is an isomorphism of vector spaces. Moreover, since (W, d_W, m_W) is a differential graded algebra, it follows that the map η is also a morphism of algebras. The acyclicity property of the injective resolutions $D_B^*(\mathbb{Z}/2)$ and $D_C^*(\mathbb{Z}/2)$ enables us to conclude that $0 \to \mathbb{Z}/2 \to W$ is an injective resolution of $\mathbb{Z}/2$. □

We here recall from [17] the Hopf algebra structure of $A = H^*(Spin(N); \mathbb{Z}/2)$:

$$A \cong \bigotimes_{3 \leq j < N,\ j\ \text{is odd}} \mathbb{Z}/2[x_j]/(x_j^{2^{n_j}}) \otimes \Lambda(z) = \Delta(x_i\ ;\ 3 \leq i < n) \otimes \Lambda(z)$$

as a Hopf algebra, where x_j is a primitive element of degree j, $\bar{\phi}(z) = \sum x_{2j} \otimes x_{2^s-2j-1}$, $\deg z = 2^s - 1$, $j2^{n_j-1} < N \leq j2^{n_j}$ and $2^{s-1} < N \leq 2^s$.

PROPOSITION 6.2. *Let L_i ($3 \leq j = 2i+1 < N$) and L be the vector subspaces of A with bases $\{x_{2i-1}, x_{2i-1}^2, \ldots, x_{2i+1}^{2^{n_j-1}}\}$ and $\{z\}$, respectively and put $L = \oplus_{3 \leq j < n} L_i$. Then*

$$0 \longrightarrow \mathbb{Z}/2 \longrightarrow A \otimes_\theta T(sL)/I = W$$

is an injective resolution of $\mathbb{Z}/2$ as an A-comodule and (W, d_W, m_W) is a differential graded algebra with the multiplication defined in Theorem 5.1.

PROOF. As in the proof of Theorem 5.4, by using the fact that $\bar{\phi}(z)$ is in $P \otimes P$, we see that $(A \otimes_\theta T(sL)/I, d_W, m_W)$ is a differential graded algebra. We define the weight of elements of W as follows:

$$\text{weight}(x_j) = \text{weight}(\theta x_j) = 0 \quad \text{and} \quad \text{weight}(z) = \text{weight}(\theta z) = 1.$$

The weight of a monomial is defined as the sum of weight of each indecomposable element. Consider the filtration of W defined by

$$F_i = \{x \in W \mid \text{weight}(x) \geq i\}.$$

The differential d_W preserves the filtration. So we have a spectral sequence converging to $H(A \otimes_\theta T(sL)/I, d_W)$ such that the E_0-term is $\sum F_i/F_{i+1}$. Since $[\theta x_j^{2^k}, \theta x_j^{2^l}] = 0$ for $i \neq j$ and $[\theta z, \theta x_j^{2^k}] = 0$ in the associated bigraded algebra

E_0, it follows that the algebra E_0 is isomorphic to $Z = A \otimes \bigotimes_j T(sL_j)/I_j \otimes T(\theta z)$ as a differential graded algebra. The multiplication of Z defines a map

$$\eta : \Gamma = \bigotimes_{3 \leq j < n} (\mathbb{Z}/2[x]/(x^{2^{n_j}}) \otimes T(sL_j)/I_j) \otimes \Lambda(z) \otimes \mathbb{Z}/2[\theta z] \to Z,$$

which is an isomorphism of differential graded modules. By virtue of Lemma 6.1, we see that Γ is acyclic and hence $H^*(E_0, d) \cong \mathbb{Z}/2$. This completes the proof. □

Let A be the Hopf algebra $H^*(Spin(N); \mathbb{Z}/2)$. We now clarify the DGA structure of

$$W' = A \square_A (A \otimes_\theta T(sL)/I) \cong A \otimes T(sL)/I$$

which gives rise to the cotorsion product $\mathrm{Cotor}_A(A, \mathbb{Z}/2)$ as an algebra. Under the notation of Proposition 6.2, let $\theta : A \to sL$ be the map which defines the twisted cochain. Then

$$\mathrm{Ker}\, \bar\theta = \mathbb{Z}/2\{x_j^{2^{n_j-1}+1}, \ldots, x_j^{2^{n_j}-1}, x_j^m x_i^n \ (i \neq j, mn \neq 0)),$$
$$x_j^m z, x_j^m x_i^n z \ (i \neq j, mn \neq 0)\}.$$

Since $\bar\theta \otimes \bar\theta \phi_A(x_j^m x_i^n z) = 0$, it follows that the ideal I is generated by $\theta x_j^{2^{n_j-1}} \theta x_j^l + \theta x_j^l \theta x_j^{2^{n_j-1}}$ ($1 \leq l \leq 2^{n_j-1} - 1$), $\theta x_j^m \theta x_i^n + \theta x_i^n \theta x_j^m$ and $\psi \bar\theta \otimes \bar\theta \phi_A(x_j^s Z)$. By virtue of Lemma 5.7, we see that the differential $d_{W'}$ on W' is given by

$$dx_j = 0 \text{ if } j \text{ is odd},$$
$$dz = \sum x_{2j} \otimes \theta x_{2^s - 2j - 1} + \sum x_{2^s - 2j - 1} \theta x_{2j},$$
$$d\theta \eta = \psi(\bar\theta \otimes \bar\theta)\phi_A \eta \text{ for } \eta \in L.$$

Lemma 5.8 gives the explicit formulae of multiplication on W' as follows:

$$\theta x_j^n \cdot x_i = x_i \theta x_j^n, \qquad \theta z \cdot x_j = x_j \theta z,$$
$$\theta x_i^n \cdot z = z\theta x_j^n + \sum x_{2^s - 2j - 1} \theta(x_{2j} x_i^n) + \sum x_{2j} \theta(x_{2^s - 2j - 1} x_i^n).$$

In particular, since

$$A = H^*(Spin(10); \mathbb{Z}/2) \cong \mathbb{Z}/2[x_3]/(x_3^4) \otimes \Lambda(x_3, x_7, x_9, z_{15})$$

as a Hopf algebra in which x_i is primitive and $\bar\phi(z_{15}) = x_3^2 \otimes x_9$, we have the following differential graded algebras $(W', d_{W'})$:

$$W' = A \otimes \mathbb{Z}/2[\theta x_3, \theta(x_3^2), \theta x_5, \theta x_7, \theta x_9, \theta z_{15}],$$

where $d_{W'}(x_i) = d_{W'}(\theta x_i) = 0$ for $i = 3, 5, 7$ and 9, $d_{W'}(z_{15}) = x_9 \theta(x_3^2) + x_3^2 \theta x_9$ and $d_{W'}(\theta z_{15}) = \theta x_3^2 \theta x_9$, which gives rise to

$$\mathrm{Cotor}_{H^*(Spin(10); \mathbb{Z}/2)}(H^*(Spin(10); \mathbb{Z}/2), \mathbb{Z}/2).$$

Observe that the algebra W' is commutative.

7. A manner for calculating the homology of a DGA

We are aware that fashions of computing the homology of differential graded algebras due to several authors are based on a certain common manner, although those computations seem to be *ad hoc* outwardly. Such examples are the computations of cotorsion products using twisted tensor products due to Kono, Mimura and Shimada in [23], [24] and [22] and that of the Hochschild-Serre spectral sequence due to Milgram and Tezuka in [29]. We mention that the computations of the Morava K-theory of Lie groups using the Atiyah-Hirzebruch spectral sequence due to Hunton, Mimura, Nishimoto and Schuster [16] [38] are also based on the manner. The purpose of this section is to describe the manner for calculating a homology.

Let (A, ε) be an augmented algebra over a field $\mathbb{K}$ with augmentation $\varepsilon : A \to \mathbb{K}$. We denote the augmented ideal $\operatorname{Ker} \varepsilon$ by A^+. Let N be an algebra without unit. Then the direct sum $A \oplus N$ of A and N is regarded as an algebra, which is an extension of A and N, equipped with the product defined by $m \cdot n = 0$ and $1 \cdot n = n$ for $m \in A^+$ and $n \in N$. Throughout this section, an algebra is supposed to be augmented unless otherwise stated.

Our manner for calculating the homology of a given differential graded algebra (A, d) is basically carried out as follows.

Step 1: We define a filtration F_* of the differential graded algebra (A, d), which makes (A, d) into a filtered differential graded algebra, and construct the spectral sequence associated with the filtration F^* converging to $H(A, d)$ as an algebra such that the E_0-term is given by

$$E_0 = \sum F_i/F_{i+1}$$

(see [7, III, 7.5], [45]). We choose the filtration, if possible, so that the images of the differential on indecomposable elements are represented by monomials in each term of the spectral sequence.

Step 2: We construct a differential graded algebra $(B \otimes C, d)$, in which $d(C) = 0$ and $d(B) \subset B \cdot C^+$, such that

$$(E_0, d_0) \cong (B \otimes C, d),$$

as a differential graded algebra. Under an appropriate filtration F_*, if possible, we arrange the algebra C and the differential d so that $C = (M \oplus N) \otimes C'$ as an algebra and $d(a) = b_a m_a$ for any $a \in B$, where M and C' are augmented algebras, N is an algebra without unit, b_a and m_a are elements of B and M, respectively. We then express such circumstance of the complex $(B \otimes C, d)$ by the following diagram consisting of axes:

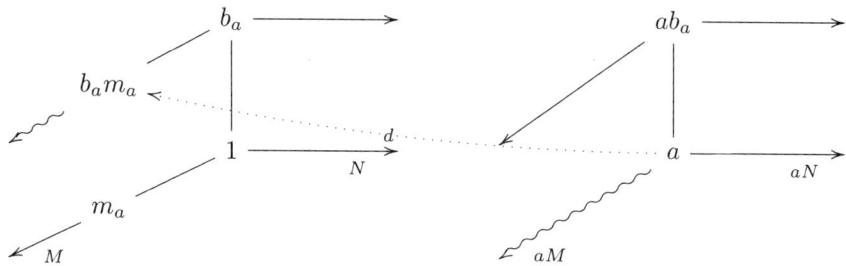

Axes drawn from an element z ($z = 1, b_a, a, ab_a$) denote the vector spaces zN and zM, respectively. The wave lines denote the vector spaces which are isomorphic each other under the differential d.

Step 3: We suppose that the E_r-term is determined as a differential graded algebra, and that all elements are represented as a point in the axes.

We calculate the homology $H(E_r, d_r)$ as a graded *module* by considering the axes; that is, we define an isomorphism of modules

$$H(E_r, d_r) \xrightarrow{\cong} \mathcal{M}$$

to a vector space $\mathcal{M}$ with a basis which is represented by elements in the axes. Observe that some elements in the axes may vanish at a previous stage.

After determining the product between cycles via the algebra structure of E_r we write explicitly an algebra structure $H(E_r, d_r)$, for example, as $(\wedge V)/I$, where $\wedge V$ is a free commutative algebra generated by a vector space V and I is an ideal of $\wedge V$. More precisely, we find indecomposable elements of $H(E_r, d_r)$ and construct a vector space with a basis consisting of the elements. A morphism of algebras $\xi : \wedge V \to H(E_r, d_r)$ is naturally defined. With the aid of the differential d_r and the module structure of $\mathcal{M}$, we find generators of an ideal I of $\wedge V$ so that the induced map

$$\widetilde{\xi} : (\wedge V)/I \to H(E_r, d_r)$$

is a well-defined morphism of algebras and the composition

$$(\wedge V)/I \xrightarrow{\widetilde{\xi}} H(E_r, d_r) \cong \mathcal{M}$$

is an isomorphism of modules. In consequence, it follows that $\widetilde{\xi}$ is an isomorphism of algebras.

Step 4: After computing the E_∞-term, we solve the extension problem on the product by representing elements in the E_∞-term by those of the original differential graded algebra (A, d).

Roughly speaking, our manner of computing a homology is the iteration of decomposition into a differential graded module and the recovery of an algebra structure.

We now illustrate more about the manner for determining a homology with computation of the homology of some differential graded algebra. The result is applied to the computation of the Hochschild homology $HH(H^*(BSpin(10); \mathbb{Z}/2))$ in Section 10.

EXAMPLE 7.1. Let $(\mathcal{K}, d)$ be a differential graded algebra over a field $\mathbb{K}$ defined by
$$\mathcal{K} = \mathbb{K}[x,y]/(xy) \otimes \Lambda(\bar{x}, \bar{y}) \otimes \Gamma[\rho], \quad d(\gamma_i(\rho)) = (\bar{x}y + \bar{y}x)\gamma_{i-1}(\rho),$$
where $\deg \bar{x} = \deg x - 1$, $\deg \bar{y} = \deg y - 1$ and $\deg \gamma_i(\rho) = \deg x + \deg y + 2$. We here compute explicitly the homology $H(\mathcal{K}, d)$ as an algebra following the procedure mentioned above. First we define the weight of elements $x, y, \bar{x}, \bar{y}$ and $\gamma_i(\rho)$ by
$$\text{weight}(x) = \text{weight}(y) = \text{weight}(\bar{x}) = \text{weight}(\gamma_i(\rho)) = 0$$
for any i and $\text{weight}(\bar{y}) = 1$. The weight of a monomial is defined as the sum of weight of the above elements. Consider the filtration of $\mathcal{K}$ defined by
$$F_i = \{z \in \mathcal{K} \mid \text{weight}(z) \geq i\}.$$
The differential d preserves the filtration. So we have a spectral sequence converging to $H(\mathcal{K})$ associated with the filtration F_* (Step 1). We see that, as an algebra,
$$E_0 \cong \Lambda(\bar{x}) \otimes \Gamma[\rho] \otimes (\mathbb{K}[y] \oplus \mathbb{K}[x]^+) \otimes \Lambda(\bar{y}) \quad \text{and} \quad d_0(\gamma_i(\rho)) = \bar{x}\gamma_{i-1}(\rho)y.$$
For simplicity, we denote $\gamma_i(\rho)$ by γ_i. We define a differential graded algebra (A, d_A) by
$$A = \Lambda(\bar{x}) \otimes \Gamma[\rho] \otimes (\mathbb{K}[y] \oplus \mathbb{K}[x]^+) \quad \text{and} \quad d_A = d_0|_A.$$
It is easily seen that
$$(E_0, d_0) \cong (A, d_A) \otimes (\Lambda(\bar{y}), 0)$$
as a differential graded algebra, and hence
$$E_1 \cong H(E_0, d_0) \cong H(A, d_A) \otimes \Lambda(\bar{y})$$
as an algebra. By using axes, the differential graded algebra (A, d_A) is expressed as follows:

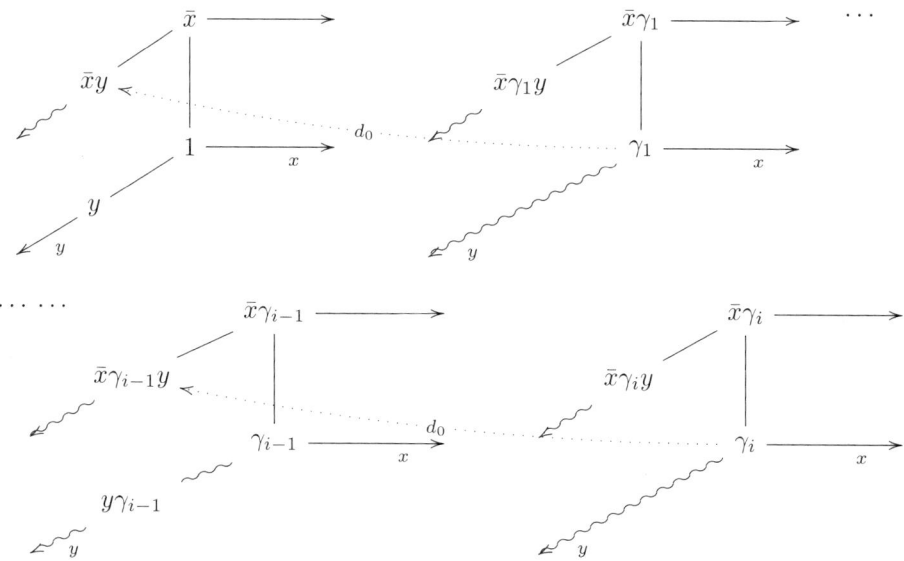

This completes Step 2. By considering the above diagram of axes, we can choose exactly a basis which forms the vector space $H(A,d)$. In consequence, we see that, as vector spaces,

$$H(A,d) \cong \mathbb{K}[y] \oplus \mathbb{K}[x]\{x, \alpha_i, \beta_j \; ; \; i \geq 1, j \geq 0\}$$
$$\cong (\mathbb{K}[y] \oplus (\mathbb{K}[x, \alpha_i \; ; \; i \geq 1]^+) \otimes \Lambda(\beta_j \; ; \; j \geq 0)/I =: \mathcal{M},$$

where $\alpha_i = x\gamma_i$, $\beta_j = \bar{x}\gamma_j$ and I is an ideal of $\mathbb{K}[x, \alpha_i] \otimes \Lambda(\beta_j)$ generated by the elements $\alpha_i\alpha_k$, $\alpha_i\beta_j$ and $\beta_j\beta_l$. The algebra structure of A enables us to conclude that the element y annihilates the elements x and α_i, and that

$$\alpha_i\alpha_k = x\gamma_i x\gamma_k = \binom{i+k}{i} x^2 \gamma_{i+k} = \binom{i+k}{i} x\alpha_{i+k},$$
$$\alpha_i\beta_j = \binom{i+j}{i} x\bar{x}\gamma_{i+j} = \binom{i+j}{i} x\beta_{i+j},$$
$$\beta_j\beta_l = 0.$$

Moreover, the element $y\beta_j$ is in Im d_A. Thus we define naturally a morphism of algebras

$$\widetilde{\xi} : \mathbb{K}[y] \oplus (\mathbb{K}[x, \alpha_i \; ; \; i \geq 1] \otimes \Lambda(\beta_j \; ; \; j \geq 0)/J)^+ \to H(A, d_A),$$

where J is the ideal generated by the elements

$$\alpha_i\alpha_k - \binom{i+k}{i} x\alpha_{i+k}, \quad \alpha_i\beta_j - \binom{i+j}{i} x\beta_{i+j}, \quad \beta_j\beta_l.$$

Comparing the module structure with that of $\mathcal{M}$, we see that the morphism $\widetilde{\xi}$ is an isomorphism of algebras. (We have here applied the argument in Step 3.) It turns

out that, as an algebra,
$$E_1 \cong \{\mathbb{K}[y] \oplus (\mathbb{K}[\alpha_i \; ; \; i \geq 0] \otimes \Lambda(\beta_j \; ; \; j \geq 0)/J)^+\} \otimes \Lambda(\bar{y}).$$

As for the differential d_1, it follows from the definition of the differential of the spectral sequence that
$$d_1(\alpha_i) = d_1(x\gamma_i) = x\bar{y}x\gamma_{i-1} = x\bar{y}\alpha_{i-1},$$
$$d_1(\beta_j) = d_i(\bar{x}\gamma_i) = \bar{x}\bar{y}x\gamma_{i-1} = -x\bar{y}\bar{x}\gamma_{i-1} = -x\bar{y}\beta_{i-1}.$$

Here we regard α_0 as the element x. The differential graded algebra (E_1, d_1) is written with axes as follows:

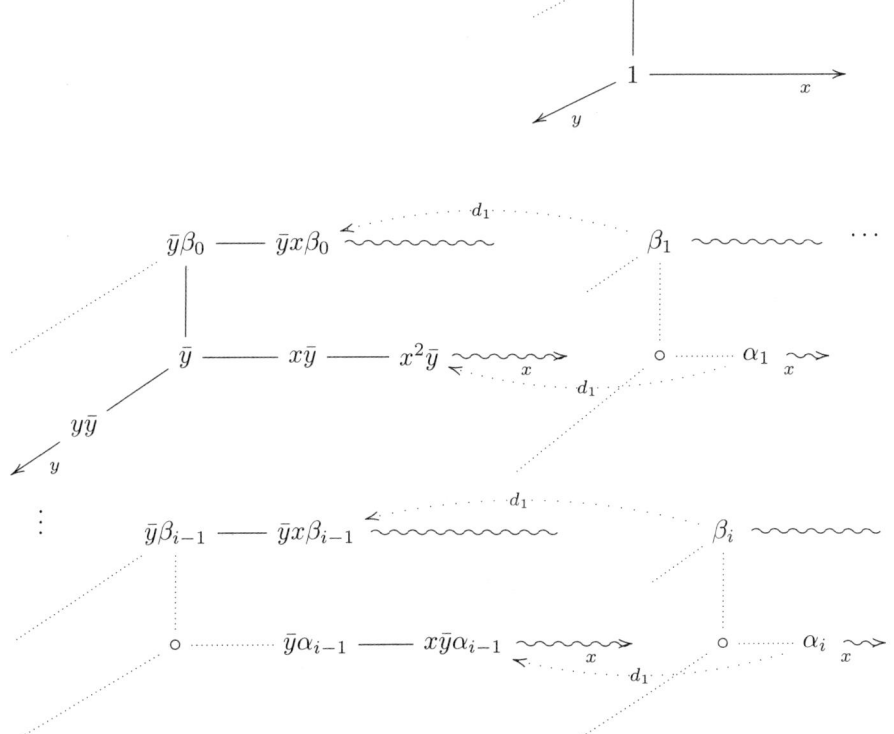

Observe that each left diagram is obtained by multiplying the previous right one by $\bar{y}$. Choosing exactly a basis which forms the E_2-term, we have
$$E_2 \cong \mathbb{K}[y]\{1, \bar{y}\} \oplus \mathbb{K}\{x, \beta_0\} \oplus \mathbb{K}\{\bar{y}x, \bar{y}\alpha_i \; ; \; i \geq 1\} \oplus \mathbb{K}\{\bar{y}\beta_j \; ; \; j \geq 0\}$$
$$\cong \frac{\mathbb{K}[y,x] \otimes \Lambda(\bar{x}, \bar{y})}{(xy, \; x^2\bar{y}, \; y\bar{x})} \oplus \mathbb{K}\{\bar{y}\alpha_i, \bar{y}\beta_j \; ; \; i, j \geq 1\} =: B$$

as vector spaces. From the algebra structure of the E_1-term, we see that $xy = 0$, $y\bar{x} = 0$, $(\bar{y}\alpha_i)z = (\bar{y}\beta_i)z = 0$ for $z = y, \bar{y}, \bar{y}\alpha_i, \bar{y}\beta_j$. Moreover, it follows that the elements $\bar{y}x^2, \bar{x}(\bar{y}\alpha_i), x^l(\bar{y}\alpha_i), \bar{x}(\bar{y}\beta_j)$ and $x^l(\bar{y}\beta_j)$ are in the image of the differential d_1 (see the above diagram of axes). Therefore we have $E_2 \cong B$ as an algebra, and hence $E_\infty \cong B$ as an algebra, since $d_s = 0$ for $s \geq 2$. We reconstruct the algebra $H(\mathcal{K}, d)$ from the algebra E_∞; that is, we solve the extension problems which appear in the E_∞-term. It is clear that the elements $x, y, \bar{x}$ and $\bar{y}$ are cycles in $\mathcal{K}$. Since $\alpha_i = x\gamma_i$ and $\beta_j = \bar{x}\gamma_j$, it follows from the definition of d that the elements $\bar{y}\alpha_i$ and $\bar{y}\alpha_i$ are also cycles in $\mathcal{K}$. We see that $y\bar{x} + x\bar{y} = 0$ in $H(\mathcal{K}, d)$, although $y\bar{x} = 0$ in the E_∞-term. If $\bar{y}\alpha_i$ and $\bar{y}\beta_j$ annihilate all the elements of degree > 0, then we can define a natural morphism of algebras

$$\eta : D := \frac{\mathbb{K}[y, x] \otimes \Lambda(\bar{x}, \bar{y})}{(xy, \ x^2\bar{y}, \ y\bar{x} + x\bar{y})} \oplus \mathbb{K}\{\bar{y}\alpha_i, \bar{y}\beta_j \ ; \ i, j \geq 1\} \to H(\mathcal{K}, d).$$

Moreover, since the composition map

$$D \xrightarrow{\eta} H(\mathcal{K}, d) = \sum F_i H(\mathcal{K}, d)/F_{i+1} H(\mathcal{K}, d) \cong E_\infty \cong M$$

is an isomorphism of modules, it follows that η is an isomorphism of algebras. The relation $x^2\bar{y} = 0$ in $H(\mathcal{K}, d)$ is deduced from the relation $y\bar{x} + x\bar{y} = 0$. We thus have

$$\frac{\mathbb{K}[y, x] \otimes \Lambda(\bar{x}, \bar{y})}{(xy, \ y\bar{x} + x\bar{y})} \oplus \mathbb{K}\{\bar{y}\alpha_i, \bar{y}\beta_j \ ; \ i, j \geq 1\} \cong H(\mathcal{K}, d)$$

as an algebra. It remains to show that the elements $\bar{y}\alpha_i$ and $\bar{y}\beta_j$ are annihilators of the ideal $H(\mathcal{K}, d)^+$. Since $\bar{y}^2 = 0$ in $H(\mathcal{K}, d)$, it suffices to prove that $\bar{y}\alpha_i$ and $\bar{y}\beta_j$ annihilate the elements x, y and $\bar{x}$ in $H(\mathcal{K}, d)$.

LEMMA 7.2. *We have the following table concerning the product on $H(\mathcal{K}, d)$.*

	$\bar{y}\alpha_i$	$\bar{y}\beta_j$
x	(1), 0	(2), 0
y	(3), 0	(4), 0
$\bar{x}$	(5), 0	(6), 0

PROOF. The results (3) and (6) follow from the algebra structure of $\mathcal{K}$. We see that

$$d(x\gamma_{i+1}) = x(y\bar{x} + x\bar{y})\gamma_i = x\bar{y}\alpha_i, \quad d(\bar{x}\gamma_{j+1}) = \bar{x}y\bar{x} + x\bar{y})\gamma_j = \bar{x}\bar{y}\beta_j$$
$$d(\bar{y}\gamma_{j+1}) = \bar{y}(y\bar{x} + x\bar{y})\gamma_j = y\bar{y}\beta_j, \quad d(\bar{x}\gamma_{i+1}) = \bar{x}(y\bar{x} + x\bar{y})\gamma_i = \bar{x}\bar{y}\alpha_i.$$

These equalities imply the results (1), (2), (4) and (5), respectively. □

This completes the computation.

REMARK 7.3. As is seen in the above example, the usage of the diagram of axes makes it possible to choose exactly a basis for $H(E_r, d_r)$, when we determine the homology as a module.

8. The Hochschild spectral sequence

We first recall the definitions of the Hochschild complex and the Connes B-map (see [**8**] and [**9**]). Let (A, d) be an augmented (not necessarily commutative) differential graded algebra with augmentation ε. The complex $(\mathbf{C}(A), b = b_0 + b_1)$ and the Connes B-map $B : \mathbf{C}(A) \to \mathbf{C}(A)$ of degree -1 are defined as follows:

$$\mathbf{C}(A) = \sum_{k=0}^{\infty} A \otimes \bar{A}^{\otimes k}, \text{ where } \bar{A} = \operatorname{Ker} \varepsilon,$$

$$b_0(\omega_0[\omega_1|\ldots|\omega_k]) = -\sum_{i=0}^{k}(-1)^{\varepsilon_{i-1}}\omega_0[\omega_1|\ldots|\omega_{i-1}|d\omega_i|\omega_{i+1}|\ldots|\omega_k],$$

$$b_1(\omega_0[\omega_1|\ldots|\omega_k]) = -\sum_{i=0}^{k-1}(-1)^{\varepsilon_i}\omega_0[\omega_1|\ldots|\omega_{i-1}|\omega_i\omega_{i+1}|\omega_{i+2}|\ldots|\omega_k]$$
$$+ (-1)^{(\deg \omega_k - 1)\varepsilon_{k-1}}\omega_k\omega_0[\omega_1|\ldots|\omega_{k-1}],$$

$$B(\omega_0[\omega_1|\ldots|\omega_k]) = \sum_{i=0}^{k}(-1)^{(\varepsilon_{i-1}+1)(\varepsilon_k-\varepsilon_{i-1})}1[\omega_i|\ldots|\omega_k|\omega_0|\ldots|\omega_{i-1}],$$

where $\deg \omega_0[\omega|\ldots|\omega_k] = \deg \omega_0 + \cdots + \deg \omega_k + k$ for $\omega_0[\omega_1|\ldots|\omega_k]$ in $\mathbf{C}(\Omega)$ and $\varepsilon_i = \deg \omega_0 + \cdots + \deg \omega_i - i$. Note that the formulae $bB + Bb = 0$ and $b^2 = B^2 = 0$ hold (see [**8**]). Observe that the bar complex BA is defined by

$$BA = \sum_{k=0}^{\infty} \bar{A}^{\otimes k}$$

with differential $d_{BA} = 1 \otimes_A b$ up to isomorphism

$$\mathbb{Z}/p \otimes_A \mathbf{C}(A) \xrightarrow{\cong} BA.$$

Define a map $\rho_A : \mathbf{C}(A) \to BA$ by

$$\rho_A(\omega_0[\omega_1|\ldots|\omega_k]) = \varepsilon(\omega_0)[\omega_1|\ldots|\omega_k].$$

It is easy to verify that ρ is a morphism of differential graded modules.

One of the reasons which allows us to calculate the HSS $\{_{HH}E_r^{*,*}, d_r\}$ is that the algebra structure of the spectral sequence is dominated by the so-called shuffle product on the Hochschild homology (see Theorem 8.1). To define the product, we first introduce the shuffle map $sh : \mathbf{C}(A) \otimes \mathbf{C}(A) \to \mathbf{C}(A \otimes A)$ defined by

$$sh(\alpha \otimes \beta) = (-1)^{\varepsilon} \sum_{\sigma:(l,m)\text{-shuffle}} (-1)^{s(\sigma)} \alpha_0 \otimes \beta_0[\xi_{\sigma(1)}|\ldots|\xi_{\sigma(l+m)}],$$

where

$$\alpha = \alpha_0[\alpha_1|\ldots|\alpha_l], \qquad \beta = \beta_0[\beta_1|\ldots|\beta_m],$$
$$(\xi_1, \ldots, \xi_{l+m}) = (\alpha_1 \otimes 1, \ldots, \alpha_l \otimes 1, 1 \otimes \beta_1, \ldots, 1 \otimes \beta_m),$$

$$\varepsilon = \deg \beta_0 (\deg \alpha_1 + \cdots + \deg \alpha_l - l),$$
$$s(\sigma) = \sum (\deg \xi_i + 1)(\deg \xi_{l+j} + 1),$$

summed over all the pairs $(i, l+j)$ with $\sigma(i) > \sigma(l+j)$, $1 \leq i \leq l$, $1 \leq j \leq m$. If (A, d) is a commutative differential graded algebra, then the multiplication m_A on A is a morphism of algebras, and hence the map induces a morphism of differential graded modules $\mathbf{C}(A \otimes A) \to \mathbf{C}(A)$. Thus we have the shuffle product

$$\widetilde{sh}_A : HH(A) \otimes HH(A) \to H(\mathbf{C}(A) \otimes \mathbf{C}(A)) \xrightarrow{H(sh)} HH(A \otimes A) \xrightarrow{HH(m_A)} HH(A).$$

Moreover, we define a product on $H(BA) = \mathrm{Tor}_A(\mathbb{Z}/p, \mathbb{Z}/p)$, which is also called the shuffle product, by

$$sh'_A : H(BA) \otimes H(BA) \to H(BA \otimes BA) \xrightarrow{H(sh)} H(B(A \otimes A)) \xrightarrow{H(B(m_A))} H(BA).$$

It is easily seen that the map ρ_A induces a morphism $H(\rho_A)$ of algebras from $HH(A)$ to $H(BA)$.

In general, we can not define a product on the Hochschild homology $HH(A, d)$ as above. However, if a given differential graded algebra (A, d) has the strongly homotopy commutative (*shc* for short) algebra structure introduced by Munkholm[**36**], it is possible to define a product on $HH(A, d)$ even if the differential graded algebra is not commutative. We describe briefly the definition of the *shc* algebra. Let

$$B : Aug\, \mathrm{DGA} \longrightarrow Coaug\, \mathrm{DGC},$$
$$\Omega : Coaug\, \mathrm{DGC} \longrightarrow Aug\, \mathrm{DGA}$$

be the bar and cobar functors, respectively*. Observe that [**36**], for any augmented differential graded algebra (A, d), there exists a quasi-isomorphism

$$\alpha_A : \Omega BA \longrightarrow A$$

such that $\alpha_A \circ \iota_A = \mathrm{id}_A$ and $\iota_A \circ \alpha_A \simeq \mathrm{id}_{\Omega BA}$ for some morphism of *differential graded modules*

$$\iota : A \longrightarrow \Omega BA.$$

An *shc* algebra is a differential graded algebra (A, d_A) equipped with the morphism of differential graded algebras

$$\mu_A : \Omega B(A \otimes A) \to \Omega BA$$

such that $\alpha_A \circ \mu_A \circ \iota_{A \otimes A} = m_A$ and satisfies the unity axiom, the associativity axiom and the commutativity axiom described in [**36**, 4.1.]. In particular, the commutativity axiom tells us that the homology of an *shc* algebra is commutative.

---

*Here $Aug\, \mathrm{DGA}$ denotes the category of the augmented differential graded algebras and $Coaug\, \mathrm{DGC}$ does the category of the coaugmented differential graded coalgebras.

Following the procedure due to Ndombol and Thomas in [**37**], we define a product on the Hochschild homology by

$$\mathbf{C}(A) \otimes \mathbf{C}(A) \xrightarrow{sh} \mathbf{C}(A \otimes A) \xleftarrow{\mathbf{C}(\alpha_{A \otimes A})} \mathbf{C}(\Omega B(A \otimes A)) \xrightarrow{\mathbf{C}(\mu)} \mathbf{C}(\Omega B A) \xrightarrow{\mathbf{C}(\alpha_A)} \mathbf{C}(A).$$

Observe that the map $\mathbf{C}(\alpha_{A \otimes A})$ induces an isomorphism on homology. The functor $\mathbf{C}(\ )$ can be replaced by the bar functor $B(\ )$ in the above sequence of maps. Therefore we can define a product on $H(BA)$, which is also called the shuffle product, by abuse of terms. We can see that the induced map

$$H(\rho_A) : HH(A) \longrightarrow H(BA)$$

becomes a morphism of algebras (see [**37**, 4.3]).

THEOREM 8.1. *Let (A, d, μ) be an shc algebra.*

(1) *There exists a spectral sequence $\{_{HH}E_r^{*,*}, d_r\}$ converging to $HH(A,d)$ as an algebra such that*

$$_{HH}E_2^{*,*} \cong HH(H(A,d))$$

as a bigraded algebra, where the product on $HH(H(A,d))$ is defined by the shuffle product associated with the commutative algebra structure of $H(A,d)$.

(2) *There exists a spectral sequence $\{_{\Omega}E_r^{*,*}, d_r\}$ converging to $HB(A,d)$ as an algebra such that*

$$_{\Omega}E_2^{*,*} \cong H(B(H(A,d))) = \mathrm{Tor}_{H(A)}(\mathbb{Z}/p, \mathbb{Z}/p)$$

as a bigraded algebra, where the product on $H(B(H(A,d)))$ is defined by the shuffle product associated with the commutative algebra structure of $H(A,d)$.

(3) *There exists a morphism $\{f_r\} : \{_{HH}E_r^{*,*}, d_r\} \to \{_{\Omega}E_r^{*,*}, d_r\}$ of spectral sequences, which induces $H(\rho) : HH(A,d) \to H(B(A,d))$, such that $f_2 = H(\rho_{H(A)})$.*

PROOF. (1) Let $\{E_r(K), d_r\}$ denote the spectral sequence constructed from a double complex (K, d_1, d_2) as usual. Since the Hochschild complex is a double complex with the external differential b_1 and the internal differential b_0, we obtain the required spectral sequence $\{_{HH}E_r^{*,*}, d_r\}$. It remains to verify that the spectral sequence has the algebra structure and that the product on the E_2-term coincides with the shuffle product on $HH(H(A,d))$ obtained from the commutative algebra structure of $H(A,d)$. Since each map defining the product on $HH(A,d)$ preserves the usual filtration of the complex which forms the spectral sequence, we have the sequence of the morphisms of spectral sequences:

$$
\begin{array}{ccccc}
E_r(\mathbf{C}(A)) \otimes E_r(\mathbf{C}(A)) & \xrightarrow{\kappa} & E_r(\mathbf{C}(A) \otimes \mathbf{C}(A)) & \xrightarrow{E_r(sh)} & E_r(\mathbf{C}(A \otimes A)) \\
& & & \nearrow\scriptstyle{E(\mathbf{C}(\alpha_{A \otimes A}))} & \\
& & \scriptstyle{\cong} & & \\
E_r(\mathbf{C}(\Omega B(A \otimes A))) & \xrightarrow[E(\mathbf{C}(\mu_A))]{} & E_r(\mathbf{C}(\Omega BA)) & \xrightarrow[E(\mathbf{C}(\alpha_A))]{} & E_r(\mathbf{C}(A)),
\end{array}
$$

whose composition is compatible with the product on the associated bigraded algebra $E_0(HH(A,d))$ when $r = \infty$, where κ is the Künneth map. The composition gives an algebra structure to the spectral sequence. Consider the case where $r = 2$. The Künneth map

$$\kappa : H(A) \otimes H(A) \longrightarrow H(A \otimes A)$$

is a morphism of algebras. Therefore, we have a commutative diagram:

$$
\begin{array}{ccccc}
E_2(\mathbf{C}(A)) \otimes E_2(\mathbf{C}(A)) & \xrightarrow{\kappa} & E_2(\mathbf{C}(A) \otimes \mathbf{C}(A)) & \xrightarrow{E_2(sh)} & E_2(\mathbf{C}(A \otimes A)) \\
\scriptstyle{\cong}\downarrow & & & & \downarrow\scriptstyle{\cong} \\
HH(H(A)) \otimes HH(H(A)) & \xrightarrow[\overline{sh}]{} & HH(H(A) \otimes H(A)) & \xrightarrow[HH(\kappa)]{} & HH(H(A \otimes A))
\end{array}
$$

where the map $\overline{sh}$ is defined by the composition

$$H(\mathbf{C}(H(A))) \otimes H(\mathbf{C}(H(A))) \to H(\mathbf{C}(H(A)) \otimes \mathbf{C}(H(A))) \xrightarrow{H(sh)} H(\mathbf{C}(H(A) \otimes H(A))).$$

Moreover, we see that the diagram

$$
\begin{array}{ccc}
E_2(\mathbf{C}(A \otimes A)) & \xleftarrow{E_2(\mathbf{C}(\alpha_{A \otimes A}))} & E_2(\mathbf{C}(\Omega(A \otimes A))) \\
\scriptstyle{\cong}\downarrow & & \downarrow\scriptstyle{\cong} \\
HH(H(A \otimes A)) & \xleftarrow{HH(H(\alpha_{A \otimes A}))} & HH(H(\Omega B(A \otimes A)))
\end{array}
$$

$$
\begin{array}{ccccc}
& \xrightarrow{E_2((\mu_A))} & E_2(\mathbf{C}(\Omega BA)) & \xrightarrow{E_2(\mathbf{C}(\alpha_A))} & E_2(\mathbf{C}(A)) \\
& & \scriptstyle{\cong}\downarrow & & \downarrow\scriptstyle{\cong} \\
& \xrightarrow{HH(H(\mu_A))} & HH(H(\Omega BA)) & \xrightarrow{HH(H(\alpha_A))} & HH(H(A))
\end{array}
$$

is commutative. Since $\alpha_A \circ \mu_A \circ \iota_{A \otimes A} = m_A$, it follows that $\alpha_A \circ \mu_A \circ \iota_{A \otimes A} \circ \alpha_{A \otimes A} = m_A \circ \alpha_{A \otimes A}$ and hence $\alpha_A \circ \mu_A \simeq m_A \circ \alpha_{A \otimes A}$ in the category of differential graded modules. Thus we see that

$$HH(H(\alpha_A)) \circ HH(H(\mu_A)) = HH(H(m_A)) \circ HH(H(\alpha_{A \otimes A})).$$

This fact yields that the composition

$$HH(H(\alpha_A)) \circ HH(H\mu_A)) \circ HH(H(\alpha_{A \otimes A}))^{-1} \circ H(\kappa) \circ \overline{sh}$$

coincides with the map

$$HH(H(m_A)) \circ HH(\kappa) \circ \overline{sh} = HH(H(m_A)\kappa) \circ \overline{sh}$$

which is the product

$$\widetilde{sh}_{H(A)} : HH(H(A)) \otimes HH(H(A)) \longrightarrow HH(H(A)).$$

This completes the proof of (1).

(2) Replacing the functor $\mathbf{C}(\ )$ by $B(\ )$, the same argument as above works well.

(3) Since the map $\rho_A : \mathbf{C}(A) \to B(A)$ preserves the filtrations which give rise to the spectral sequences in (1) and (2), it follows that the map induces the required morphism of spectral sequences. □

Since the normalized cochain algebra $C^*(X)$ for a simply connected space X possesses an *shc* algebra structure [**36**], combining [**37**, Theorem 2] with Proposition 8.1 (1), we have the first spectral sequence HSS mentioned in Introduction. From Proposition 8.1 (2), (3) and the latter half of [**37**, Theorem 2], we can establish the following theorem.

THEOREM 8.2. *Let X be a simply connected space. There exist a spectral sequence $\{_\Omega E_r^{*,*}, d_r\}$ converging to $H^*(\Omega X; \mathbb{Z}/p)$ as an algebra and a morphism of spectral sequences from the HSS $\{_{HH}E_r^{*,*}, d_r\}$ converging to $H^*(X^{S^1}; \mathbb{Z}/p)$ to $\{_\Omega E_r^{*,*}, d_r\}$ such that*

(i) $_\Omega E_2^{*,*} \cong H(B(H^*(X; \mathbb{Z}/p))) = \mathrm{Tor}_{H^*(X;\mathbb{Z}/p)}(\mathbb{Z}/p, \mathbb{Z}/p)$ *as an algebra;*

(ii) *the map*

$$j^* : H^*(X^{S^1}; \mathbb{Z}/p) \to H^*(\Omega X; \mathbb{Z}/p)$$

induced from the inclusion $j : \Omega X \to X^{S^1}$ preserves the filtration which is given by the spectral sequences and hence the map

$$E_0(j^*) : E_0 H^*(X^{S^1}; \mathbb{Z}/p) \to E_0 H^*(\Omega X; \mathbb{Z}/p)$$

between the associated bigraded algebras can be defined;

(iii) *the map f_∞ is compatible with the induced map j^*; that is, f_∞ coincides with $E_0(j^*)$ up to isomorphisms*

$$_{HH}E_\infty^{*,*} \cong E_0 H^*(X^{S^1}; \mathbb{Z}/p),$$
$$_\Omega E_\infty^{*,*} \cong E_0 H^*(\Omega X; \mathbb{Z}/p).$$

9. Proof of Theorem 1.6

In this section, A and B denote the Hopf algebra $H^*(PU(3);\mathbb{Z}/3)$ and the cohomology algebra

$$H^*(BPU(3);\mathbb{Z}/3) = \mathbb{Z}/3[y_2, y_8, y_{12}] \otimes \Lambda(y_3, y_7)/(y_2 y_3, y_2 y_7, y_3 y_7 + y_2 y_8),$$

respectively.

THEOREM 9.1. *As an* $H^*(BPU(3);\mathbb{Z}/3)$-*algebra*,

$$\mathrm{Cotor}_{H^*(PU(3);\mathbb{Z}/3)}(H^*(PU(3);\mathbb{Z}/3), \mathbb{Z}/3)$$
$$\cong \mathbb{Z}/3[x_2, y_3, z_6, y_8, z_8, y_{12}] \otimes \Lambda(x_1, y_3, y_7, z_9, z_{11})/I,$$

in which

$$y_{i+1} = \{\theta x_i\}, \quad \text{for } i = 1, 2,$$
$$y_7 = \{\theta x_2 \theta x_3 - \theta x_1 \theta(x_2^2)\},$$
$$y_8 = \{\theta x_2 \theta(x_2^2) + \theta(x_2^2) \theta x_2\},$$
$$y_{12} = \{(\theta(x_3))^3\},$$
$$z_6 = \{x_3 \theta x_2 + x_2 \theta x_3 + x_1 \theta(x_2^2)\},$$
$$z_8 = \{x_1 x_2^2 x_3\},$$
$$z_9 = \{x_2^2 x_3 \theta x_1 - x_1 x_2^2 \theta x_3\},$$
$$z_{11} = \{x_1 x_3 (\theta x_2 \theta x_3 - \theta x_1 \theta(x_2^2)) - x_1 x_2^2 (\theta x_3)^2\},$$

where $\{u\}$ *denotes the element in the cotorsion product represented by a cocycle* u *in* W'. *Here* I *is the ideal generated by elements*

$x_2 y_2 + x_1 y_3,$	$y_2 y_3,$
$x_2^3,$	$y_2 z_6 + x_1 y_7,$
$x_1 z_8,$	$y_2 y_7$
$z_6 y_3 - x_2 y_7 - x_1 y_8,$	$x_1 z_9 - y_2 z_8,$
$x_2 z_8,$	$y_2 y_8 + y_3 y_7,$
$z_8 y_3 - x_1 x_2^2 z_6,$	$x_1 z_{11},$
$x_2 z_9,$	$z_9 y_3 - x_1 x_2^2 y_7,$
$x_1 z_6^2 + x_2 z_{11},$	$y_2 z_{11},$
$z_{11} y_3 + x_1 z_6 y_7,$	$z_6 z_8,$
$z_6 z_9 + z_8 y_7,$	$z_8 y_7 - x_2^2 z_{11},$

$$x_2^2 z_6^2 - z_8 y_8, \qquad\qquad z_9 y_7,$$

$$z_8^2, \qquad\qquad x_2^2 z_6 y_7 + z_9 y_8,$$

$$z_6 z_{11} + x_1 x_2^2 y_{12}, \qquad\qquad z_8 z_9,$$

$$z_6^3, \qquad\qquad z_{11} y_7 + x_1 x_2 y_3 y_{12},$$

$$-z_6^2 y_7 + z_{11} y_8 + x_2^2 y_3 y_{12}, \qquad\qquad z_8 z_{11},$$

$$z_9 z_{11},$$

PROOF. We denote $\theta x_1, \theta x_2, \theta x_3$ and θx_2^2 by a_1, a_2, b_4 and c_5 respectively. We define a filtration of the complex $(W', d_{W'})$ by

$$F_i = \{ x \otimes c \in A \otimes \overline{X} \mid |c| \geq i \}.$$

Consider the spectral sequence associated with the filtration converging to $\operatorname{Cotor}_A(A, \mathbb{Z}/3)$. The E_1-term is $A \otimes \overline{X}$ with differential

$$d_1(x_i) = 0, \quad d_1(a_i) = 0, \quad d_1(b_4) = -a_2 a_3, \quad d_1(c_5) = a_3^2.$$

Then the E_2-term is isomorphic to $A \otimes \operatorname{Cotor}_A(\mathbb{Z}/3, \mathbb{Z}/3)$ as an algebra. We have

$$\operatorname{Cotor}_A(\mathbb{Z}/3, \mathbb{Z}/3) = \mathbb{Z}/3[a_2, e_8, e_{12}] \otimes \Lambda(a_3, e_7)/(a_2 a_3, a_2 e_7, a_3 e_7 + a_2 e_8),$$

where e_7, e_8 and e_{12} are represented by $a_3 b_4 - a_2 c_5$, $a_3 c_5 + c_5 a_3$ and b_4^3 respectively. The first non-zero differential is given by

$$d_2(x_3) = x_2 a_2.$$

In order to calculate the E_3-term, we consider the following differential graded module:

$$(M, d) = (\mathbb{Z}/3[x_2]/(x_2^3) \otimes \Lambda(x_3) \otimes (\mathbb{Z}/3[e_8]\{1, a_2\} \oplus \mathbb{Z}/3[a_2]\{a_2^2\}, d),$$

where $d(x_3) = x_2 a_2$.

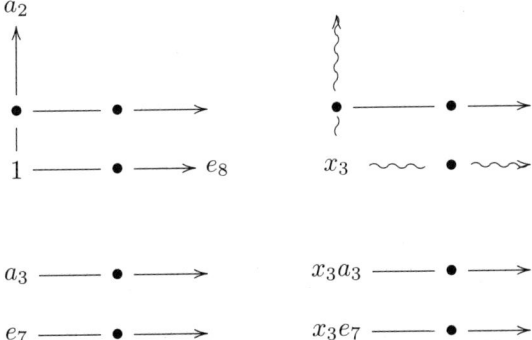

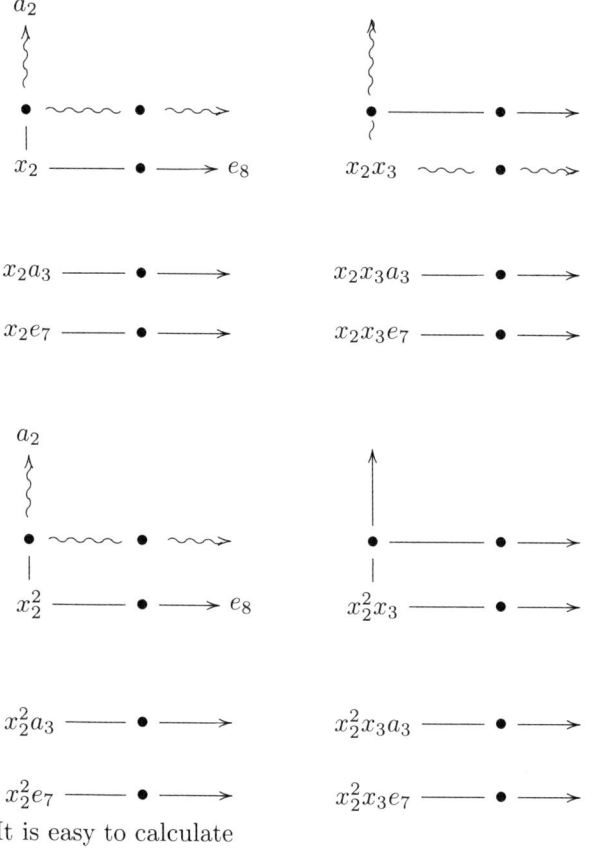

It is easy to calculate

$$H(M) = \mathbb{Z}/3[e_8] \otimes \mathbb{Z}/3[x_2]/(x_2^3)\{1, a_3, e_7, f_6, f_{10}, f_6e_7\}$$
$$\oplus \mathbb{Z}/3[e_8]\{a_2e_8, f_7\} \oplus \mathbb{Z}/3[a_2]\{a_2, a_2f_7\},$$

where $f_6 = x_3a_3 + x_2b_4 + x_1c_5$, $f_7 = x_2^2 x_3$ and $f_{10} = x_3(a_3b_4 - a_2c_5) - x_2b_4^2 + x_1c_5b_4$. We write

$$(E_2, d_2) = (\Lambda(x_1) \otimes \mathbb{Z}/3[e_{12}], 0) \otimes (M, d).$$

Thus we have

$$E_3 = \Lambda(x_1) \otimes \mathbb{Z}/3[e_{12}] \otimes (\mathbb{Z}/3[e_8] \otimes \mathbb{Z}/3[x_2]/(x_2^3)\{1, a_3, e_7, f_6, f_{10}, f_6e_7\}$$
$$\oplus \mathbb{Z}/3[e_8]\{a_2e_8, f_7\} \oplus \mathbb{Z}/3[a_2]\{a_2, a_2f_7\}).$$

LEMMA 9.2. *The algebra structure of E_3 is given by*

$$\mathbb{Z}/3[x_2] \otimes \Lambda(x_1) \otimes \mathbb{Z}/3[a_2, e_8, e_{12}, f_6, f_{10}] \otimes \Lambda(a_3, e_7, f_7)/I,$$

where I is generated by

$$I = (x_2a_2, a_2a_3, x_2^3, a_2f_6, a_2e_7, x_2f_7, f_6a_3, a_3e_7 + a_2e_8, f_7a_3 - x_2^2f_6,$$
$$a_2f_{10}, f_6^2, f_{10}a_3 + f_6e_7, f_6f_7, f_7e_7 - x_2^2f_{10}, f_6f_{10}, f_{10}e_7, f_7f_{10}, f_{10}^2).$$

9. PROOF OF THEOREM 1.6

PROOF. First we check that generators of I are zero in the E_3-term. It is easy to see that x_2^3 and $x_2 f_7$ are zero in the chain level, that $a_2 a_3$, $a_2 f_6$, $a_2 e_7$, $f_6 a_3$, $a_3 e_7 + a_2 e_8$, $f_7 a_3 - x_2^2 f_6$, $a_2 f_{10}$, f_6^2, $f_{10} a_3 + f_6 e_7$, $f_6 f_7$, $f_7 e_7 - x_2^2 f_{10}$, $f_6 f_{10}$, $f_{10} e_7$, $f_7 f_{10}$ and f_{10}^2 are zero in the E_2-term and that $x_2 a_2$ is in Im d_2. In order to check that there are no more relations, it is sufficient to show that, using the relations, the action of the product of the generators of E_3 are closed in the module structure. It is obvious that the action of the product of x_1 and e_{12} are closed. The following table is that of the other generators.

	x_2	a_2	a_3	e_7
e_8^i ($i \geq 0$)	$x_2 e_8^i$	$a_2 e_8^i$	$a_3 e_8^i$	$e_7 e_8^i$
$x_2 e_8^i$ ($i \geq 0$)	$x_2^2 e_8^i$	0	$x_2 a_3 e_8^i$	$x_2 e_7 e_8^i$
$x_2^2 e_8^i$ ($i \geq 0$)	0	0	$x_2^2 a_3 e_8^i$	$x_2^2 e_7 e_8^i$
a_2	0	a_2^2	0	0
$a_2 e_8^i$ ($i \geq 1$)	0	0	0	0
a_2^i ($i \geq 2$)	0	a_2^{i+1}	0	0
$a_3 e_8^i$ ($i \geq 0$)	$x_2 a_3 e_8^i$	0	0	$-a_2 e_8^{i+1}$
$x_2 a_3 e_8^i$ ($i \geq 0$)	$x_2^2 a_3 e_8^i$	0	0	0
$x_2^2 a_3 e_8^i$ ($i \geq 0$)	0	0	0	0
$e_7 e_8^i$ ($i \geq 0$)	$x_2 e_7 e_8^i$	0	$a_2 e_8^{i+1}$	0
$x_2 e_7 e_8^i$ ($i \geq 0$)	$x_2^2 e_7 e_8^i$	0	0	0
$x_2^2 e_7 e_8^i$ ($i \geq 0$)	0	0	0	0
$f_6 e_8^i$ ($i \geq 0$)	$x_2 f_6 e_8^i$	0	0	$f_6 e_7 e_8^i$
$x_2 f_6 e_8^i$ ($i \geq 0$)	$x_2^2 f_6 e_8^i$	0	0	$x_2 f_6 e_7 e_8^i$
$x_2^2 f_6 e_8^i$ ($i \geq 0$)	0	0	0	$x_2^2 f_6 e_7 e_8^i$
f_7	0	$a_2 f_7$	$x_2^2 f_6$	$x_2^2 f_{10}$
$f_7 e_8^i$ ($i \geq 1$)	0	$-x_2^2 f_6 e_7 e_8^{i-1}$	$x_2^2 f_6 e_8^i$	$x_2^2 f_{10} e_8^i$
$a_2 f_7$	0	$a_2^2 f_7$	0	0
$a_2^i f_7$ ($i \geq 2$)	0	$a_2^{i+1} f_7$	0	0
$f_{10} e_8^i$ ($i \geq 0$)	$x_2 f_{10} e_8^i$	0	$-f_6 e_7 e_8^i$	0
$x_2 f_{10} e_8^i$ ($i \geq 0$)	$x_2^2 f_{10} e_8^i$	0	$-x_2 f_6 e_7 e_8^i$	0
$x_2^2 f_{10} e_8^i$ ($i \geq 0$)	0	0	$-x_2^2 f_6 e_7 e_8^i$	0
$f_6 e_7 e_8^i$ ($i \geq 0$)	$x_2 f_6 e_7 e_8^i$	0	0	0
$x_2 f_6 e_7 e_8^i$ ($i \geq 0$)	$x_2^2 f_6 e_7 e_8^i$	0	0	0
$x_2^2 f_6 e_7 e_8^i$ ($i \geq 0$)	0	0	0	0

	e_8	f_6	f_7	f_{10}
e_8^i ($i \geq 0$)	e_8^{i+1}	$f_6 e_8^i$	$f_7 e_8^i$	$f_{10} e_8^i$
$x_2 e_8^i$ ($i \geq 0$)	$x_2 e_8^{i+1}$	$x_2 f_6 e_8^i$	0	$x_2 f_{10} e_8^i$
$x_2^2 e_8^i$ ($i \geq 0$)	$x_2^2 e_8^{i+1}$	$x_2^2 f_6 e_8^i$	0	$x_2^2 f_{10} e_8^i$
a_2	$a_2 e_8$	0	$a_2 f_7$	0
$a_2 e_8^i$ ($i \geq 1$)	$a_2 e_8^{i+1}$	0	$a_2 f_7 e_8^i$	0
a_2^i ($i \geq 2$)	0	0	$a_2^i f_7$	0
$a_3 e_8^i$ ($i \geq 0$)	$a_3 e_8^{i+1}$	0	$-x_2^2 f_6 e_8^i$	$-f_6 e_7 e_8^i$
$x_2 a_3 e_8^i$ ($i \geq 0$)	$x_2 a_3 e_8^{i+1}$	0	0	$-x_2 f_6 e_7 e_8^i$
$x_2^2 a_3 e_8^i$ ($i \geq 0$)	$x_2^2 a_3 e_8^{i+1}$	0	0	$-x_2^2 f_6 e_7 e_8^i$
$e_7 e_8^i$ ($i \geq 0$)	$e_7 e_8^{i+1}$	$f_6 e_7 e_8^i$	$-x_2^2 f_{10} e_8^i$	0
$x_2 e_7 e_8^i$ ($i \geq 0$)	$x_2 e_7 e_8^{i+1}$	$x_2 f_6 e_7 e_8^i$	0	0
$x_2^2 e_7 e_8^i$ ($i \geq 0$)	$x_2^2 e_7 e_8^{i+1}$	$x_2^2 f_6 e_7 e_8^i$	0	0
$f_6 e_8^i$ ($i \geq 0$)	$f_6 e_8^{i+1}$	0	0	0
$x_2 f_6 e_8^i$ ($i \geq 0$)	$x_2 f_6 e_8^{i+1}$	0	0	0
$x_2^2 f_6 e_8^i$ ($i \geq 0$)	$x_2^2 f_6 e_8^{i+1}$	0	0	0
f_7	$f_7 e_8$	0	0	0
$f_7 e_8^i$ ($i \geq 1$)	$f_7 e_8^{i+1}$	0	0	0
$a_2 f_7$	$-x_2^2 f_6 e_7$	0	0	0
$a_2^i f_7$ ($i \geq 2$)	0	0	0	0
$f_{10} e_8^i$ ($i \geq 0$)	$f_{10} e_8^{i+1}$	0	0	0
$x_2 f_{10} e_8^i$ ($i \geq 0$)	$x_2 f_{10} e_8^{i+1}$	0	0	0
$x_2^2 f_{10} e_8^i$ ($i \geq 0$)	$x_2^2 f_{10} e_8^{i+1}$	0	0	0
$f_6 e_7 e_8^i$ ($i \geq 0$)	$f_6 e_7 e_8^{i+1}$	0	0	0
$x_2 f_6 e_7 e_8^i$ ($i \geq 0$)	$x_2 f_6 e_7 e_8^{i+1}$	0	0	0
$x_2^2 f_6 e_7 e_8^i$ ($i \geq 0$)	$x_2^2 f_6 e_7 e_8^{i+1}$	0	0	0

This completes the proof of the lemma. □

The next non-zero differentials appear in the E_3-term and are given by $d_3(f_7) = x_1 x_2^2 a_3$ and $d_3(f_{10}) = -x_1 a_2 e_8$. In order to calculate the E_4-term, we consider the following differential graded module:

$$(M, d) = (\Lambda(x_1) \otimes (\mathbb{Z}/3[e_8] \otimes \mathbb{Z}/3[x_2]/(x_2^3)\{1, a_3, e_7, f_6, f_{10}, f_6 e_7\}$$
$$\oplus \mathbb{Z}/3[e_8]\{a_2 e_8, f_7\} \oplus \mathbb{Z}/3[a_2]\{a_2, a_2 f_7\}, d),$$

where $d(f_7) = x_1 x_2^2 a_3$ and $d(f_{10}) = -x_1 a_2 e_8$.

9. PROOF OF THEOREM 1.6

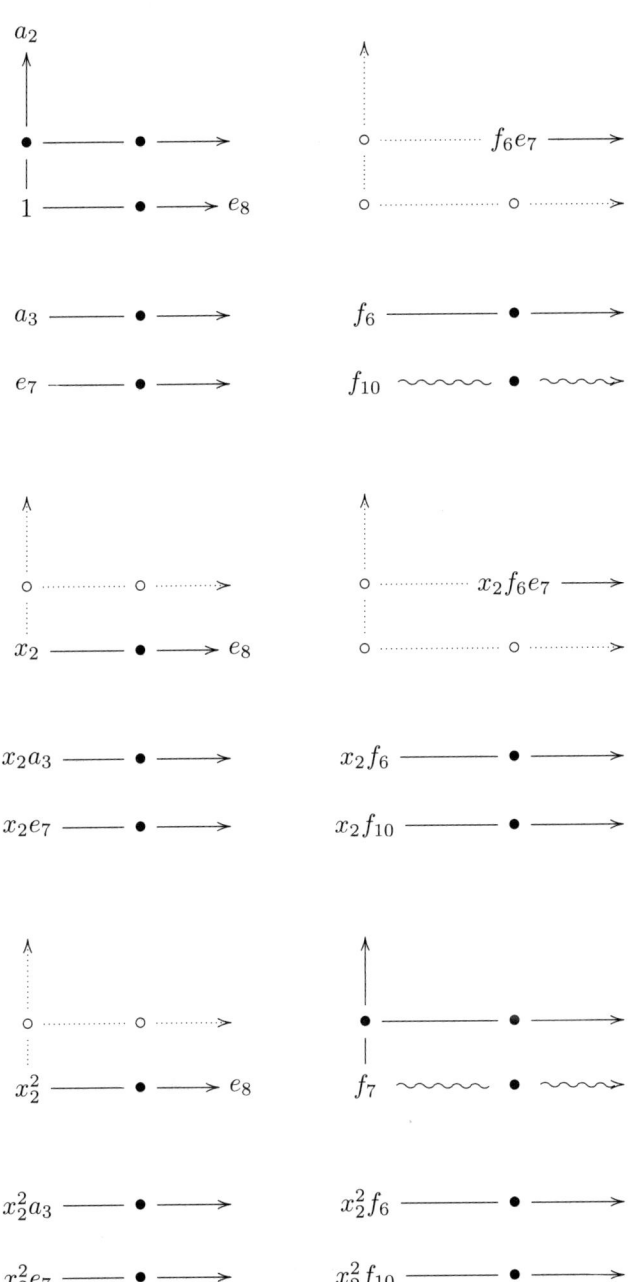

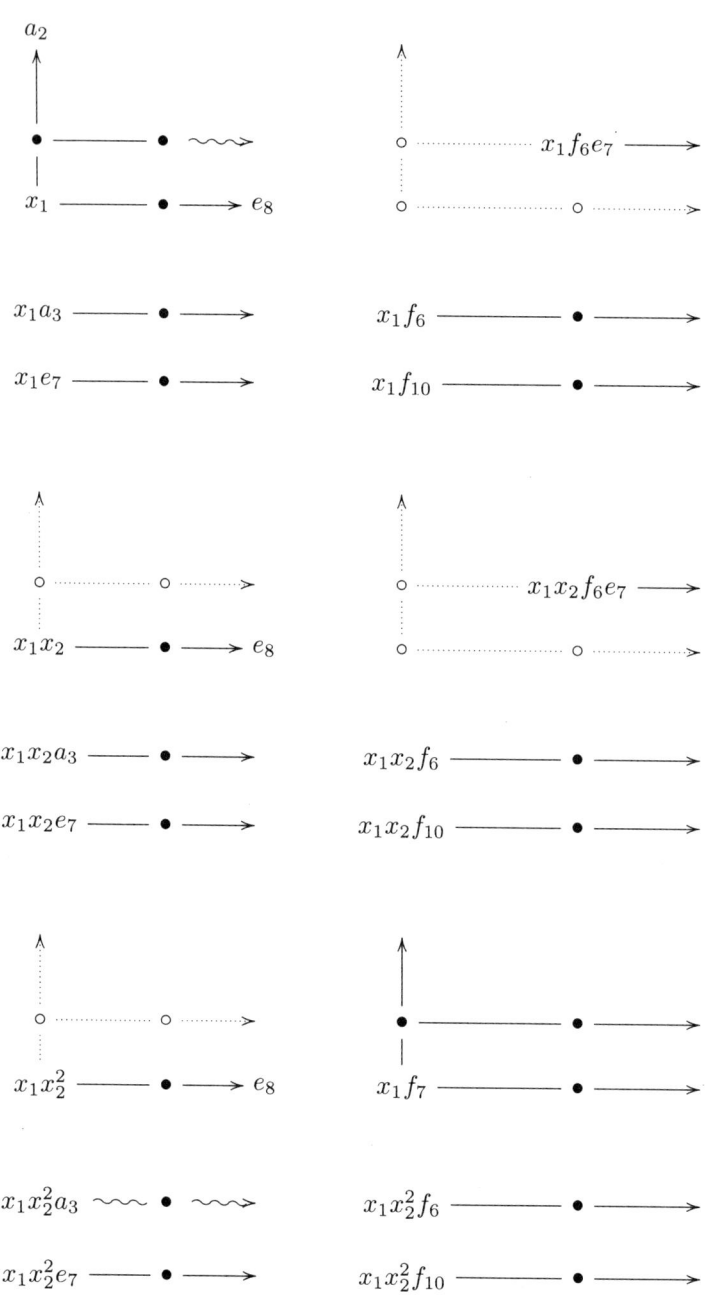

9. PROOF OF THEOREM 1.6

It is easy to calculate

$$H(M) = \mathbb{Z}/3[e_8] \otimes \mathbb{Z}/3[x_2]/(x_2^3)\{1, a_3, e_7, f_6, f_6e_7, x_1, x_1e_7, x_1f_6, x_1f_6e_7, f_{11}\}$$
$$\oplus \mathbb{Z}/3[e_8]\{a_2e_8, f_{12}, x_2f_{12}, x_1a_3, x_1x_2a_3, f_8\}$$
$$\oplus \mathbb{Z}/3[a_2]\{a_2, f_9, x_1a_2, a_2f_8\},$$

where $f_8 = x_1x_2^2x_3$, $f_9 = x_2^2x_3a_2 - x_1x_2^2b_4$, $f_{11} = x_1x_3(a_3b_4 - a_2c_5) - x_1x_2b_4^2$, $f_{12} = x_2x_3(a_3b_4 - a_2c_5) - x_2^2b_4^2 + x_1x_2c_5b_4 - x_1x_3(a_3c_5 + c_5a_3)$. Since we have

$$(E_3, d_3) = (\mathbb{Z}/3[e_{12}], 0) \otimes (M, d),$$

we obtain

$$E_4 = \mathbb{Z}/3[e_{12}] \otimes (\mathbb{Z}/3[e_8] \otimes \mathbb{Z}/3[x_2]/(x_2^3)\{1, a_3, e_7, f_6, f_6e_7, x_1,$$
$$x_1e_7, x_1f_6, x_1f_6e_7, f_{11}\}$$
$$\oplus \mathbb{Z}/3[e_8]\{a_2e_8, f_{12}, x_2f_{12}, x_1a_3, x_1x_2a_3, f_8\}$$
$$\oplus \mathbb{Z}/3[a_2]\{a_2, f_9, x_1a_2, a_2f_8\}).$$

LEMMA 9.3. *The algebra structure of E_4 is given by*

$$\Lambda(x_1) \otimes \mathbb{Z}/3[x_2] \otimes \Lambda(a_3, e_7, f_9, f_{11}) \otimes \mathbb{Z}/3[a_2, f_6, e_8, f_8, e_{12}, f_{12}]/I,$$

where I is generated by

$$I = (x_2a_2, a_2a_3, x_2^3, x_1x_2^2a_3, a_2f_6, x_1f_8, a_2e_7, f_6a_3, x_1f_9 - a_2f_8, x_2f_8, a_2e_8 + a_3e_7,$$
$$x_1a_2e_8, f_8a_3 - x_1x_2^2f_6, x_1f_{11}, x_2f_9, f_9a_3, f_6^2, x_1f_{12} - x_2f_{11}, a_2f_{11}, a_2f_{12},$$
$$f_{11}a_3 + x_1f_6e_7, f_6f_8, f_{12}a_3 + x_2f_6e_7, f_6f_9, f_8e_7 - x_2^2f_{11}, x_2^2f_{12}, f_9e_7, f_8^2,$$
$$x_2^2f_6e_7 + f_9e_8, f_6f_{11}, f_8f_9, f_6f_{12}, f_{11}e_7, f_{12}e_7, f_8f_{11}, f_8f_{12}, f_9f_{11}, f_9f_{12},$$
$$f_{11}f_{12}, f_{12}^2).$$

PROOF. First we check that generators of I are zero in the E_4-term. It is obvious that the generators except for $x_1x_2^2a_3$ and $x_1a_2e_8$ are zero in the E_3-term and that $x_1x_2^2a_3$ and $x_1a_2e_8$ are in Im d_3. In order to check that there are no more relations, it is sufficient to show that, using the relations, the action of the product of the generators of E_4 are closed in the module structure. It is obvious that the action of the product of e_{12} are closed. The following table is that of the other generators.

	x_1	x_2	a_2	a_3	f_6
e_8^i $(i \geq 0)$	$x_1 e_8^i$	$x_2 e_8^i$	$a_2 e_8^i$	$a_3 e_8^i$	$f_6 e_8^i$
$x_2 e_8^i$ $(i \geq 0)$	$x_1 x_2 e_8^i$	$x_2^2 e_8^i$	0	$x_2 a_3 e_8^i$	$x_2 f_6 e_8^i$
$x_2^2 e_8^i$ $(i \geq 0)$	$x_1 x_2^2 e_8^i$	0	0	$x_2^2 a_3 e_8^i$	$x_2^2 f_6 e_8^i$
a_2	$x_1 a_2$	0	a_2^2	0	0
$a_2 e_8^i$ $(i \geq 1)$	0	0	0	0	0
a_2^i $(i \geq 2)$	$x_1 a_2^i$	0	a_2^{i+1}	0	0
$a_3 e_8^i$ $(i \geq 0)$	$-x_1 a_3 e_8^i$	$x_2 a_3 e_8^i$	0	0	0
$x_2 a_3 e_8^i$ $(i \geq 0)$	$-x_1 x_2 a_3 e_8^i$	$x_2^2 a_3 e_8^i$	0	0	0
$x_2^2 a_3 e_8^i$ $(i \geq 0)$	0	0	0	0	0
$e_7 e_8^i$ $(i \geq 0)$	$-x_1 e_7 e_8^i$	$x_2 e_7 e_8^i$	0	$a_2 e_8^{i+1}$	$f_6 e_7 e_8^i$
$x_2 e_7 e_8^i$ $(i \geq 0)$	$-x_1 x_2 e_7 e_8^i$	$x_2^2 e_7 e_8^i$	0	0	$x_2 f_6 e_7 e_8^i$
$x_2^2 e_7 e_8^i$ $(i \geq 0)$	$-x_1 x_2 e_7 e_8^i$	0	0	0	$x_2^2 f_6 e_7 e_8^i$
$f_6 e_8^i$ $(i \geq 0)$	$x_1 f_6 e_8^i$	$x_2 f_6 e_8^i$	0	0	0
$x_2 f_6 e_8^i$ $(i \geq 0)$	$x_1 x_2 f_6 e_8^i$	$x_2^2 f_6 e_8^i$	0	0	0
$x_2^2 f_6 e_8^i$ $(i \geq 0)$	$x_1 x_2^2 f_6 e_8^i$	0	0	0	0
f_9	$-a_2 f_8$	0	$a_2 f_9$	0	0
$f_9 a_2^i$ $(i \geq 1)$	$-a_2^{i+1} f_8$	0	$a_2^{i+1} f_9$	0	0
$f_{12} e_8^i$ $(i \geq 0)$	$x_2 f_{11} e_8^i$	$x_2 f_{12} e_8^i$	0	$-x_2 f_6 e_7 e_8^i$	0
$x_2 f_{12} e_8^i$ $(i \geq 0)$	$x_2^2 f_{11} e_8^i$	0	0	$-x_2^2 f_6 e_7 e_8^i$	0
$f_6 e_7 e_8^i$ $(i \geq 0)$	$-x_1 f_6 e_7 e_8^i$	$x_2 f_6 e_7 e_8^i$	0	0	0
$x_2 f_6 e_7 e_8^i$ $(i \geq 0)$	$-x_1 x_2 f_6 e_7 e_8^i$	$x_2^2 f_6 e_7 e_8^i$	0	0	0
$x_2^2 f_6 e_7 e_8^i$ $(i \geq 0)$	$-x_1 x_2^2 f_6 e_7 e_8^i$	0	0	0	0
x_1	0	$x_1 x_2$	$x_1 a_2$	$x_1 a_3$	$x_1 f_6$
$x_1 e_8^i$ $(i \geq 1)$	0	$x_1 x_2 e_8^i$	0	$x_1 a_3 e_8^i$	$x_1 f_6 e_8^i$
$x_1 x_2 e_8^i$ $(i \geq 0)$	0	$x_1 x_2^2 e_8^i$	0	$x_1 x_2 a_3 e_8^i$	$x_1 x_2 f_6 e_8^i$
$x_1 x_2^2 e_8^i$ $(i \geq 0)$	0	0	0	$x_1 x_2^2 a_3 e_8^i$	$x_1 x_2^2 f_6 e_8^i$
$x_1 a_2^i$ $(i \geq 1)$	0	0	$x_1 a_2^{i+1}$	0	0
$x_1 a_3 e_8^i$ $(i \geq 0)$	0	$x_1 x_2 a_3 e_8^i$	0	0	0
$x_1 x_2 a_3 e_8^i$ $(i \geq 0)$	0	0	0	0	0
$x_1 e_7 e_8^i$ $(i \geq 0)$	0	$x_1 x_2 e_7 e_8^i$	0	0	$x_1 f_6 e_7 e_8^i$
$x_1 x_2 e_7 e_8^i$ $(i \geq 0)$	0	$x_1 x_2^2 e_7 e_8^i$	0	0	$x_1 x_2 f_6 e_7 e_8^i$
$x_1 x_2^2 e_7 e_8^i$ $(i \geq 0)$	0	0	0	0	$x_1 x_2^2 f_6 e_7 e_8^i$
$x_1 f_6 e_8^i$ $(i \geq 0)$	0	$x_1 x_2 f_6 e_8^i$	0	0	0
$x_1 x_2 f_6 e_8^i$ $(i \geq 0)$	0	$x_1 x_2^2 f_6 e_8^i$	0	0	0

9. PROOF OF THEOREM 1.6

	x_1	x_2	a_2	a_3	f_6
$x_1 x_2^2 f_6 e_8^i$ $(i \geq 0)$	0	0	0	0	0
$f_8 e_8^i$ $(i \geq 0)$	0	0	$a_2 f_8 e_8^i$	$x_1 x_2^2 f_6 e_8^i$	0
$a_2 f_8$	0	0	$a_2^2 f_8$	0	0
$a_2^i f_8$ $(i \geq 2)$	0	0	$a_2^{i+1} f_8$	0	0
$f_{11} e_8^i$ $(i \geq 0)$	0	$x_2 f_{11} e_8^i$	0	$x_1 f_6 e_7 e_8^i$	0
$x_2 f_{11} e_8^i$ $(i \geq 0)$	0	$x_2^2 f_{11} e_8^i$	0	$x_1 x_2 f_6 e_7 e_8^i$	0
$x_2^2 f_{11} e_8^i$ $(i \geq 0)$	0	0	0	$x_1 x_2^2 f_6 e_7 e_8^i$	0
$x_1 f_6 e_7 e_8^i$ $(i \geq 0)$	0	$x_1 x_2 f_6 e_7 e_8^i$	0	0	0
$x_1 x_2 f_6 e_7 e_8^i$ $(i \geq 0)$	0	$x_1 x_2^2 f_6 e_7 e_8^i$	0	0	0
$x_1 x_2^2 f_6 e_7 e_8^i$ $(i \geq 0)$	0	0	0	0	0

	e_7	e_8	f_8	f_9	f_{11}
e_8^i $(i \geq 1)$	$e_7 e_8^i$	e_8^{i+1}	$f_8 e_8^i$	$-x_2^2 f_6 e_7 e_8^{i-1}$	$f_{11} e_8^i$
$x_2 e_8^i$ $(i \geq 0)$	$x_2 e_7 e_8^i$	$x_2 e_8^{i+1}$	0	0	$x_2 f_{11} e_8^i$
$x_2^2 e_8^i$ $(i \geq 0)$	$x_2^2 e_7 e_8^i$	$x_2^2 e_8^{i+1}$	0	0	$x_2^2 f_{11} e_8^i$
a_2	0	$a_2 e_8$	$a_2 f_8$	$a_2 f_9$	0
$a_2 e_8^i$ $(i \geq 1)$	0	$a_2 e_8^{i+1}$	$-x_1 x_2^2 f_6 e_7 e_8^{i-1}$	0	0
a_2^i $(i \geq 2)$	0	0	$a_2^i f_8$	$a_2^i f_9$	0
$a_3 e_8^i$ $(i \geq 0)$	$-a_2 e_8^{i+1}$	$a_3 e_8^{i+1}$	$x_1 x_2^2 f_6 e_8^i$	0	$x_1 f_6 e_7 e_8^i$
$x_2 a_3 e_8^i$ $(i \geq 0)$	0	$x_2 a_3 e_8^{i+1}$	0	0	$x_1 x_2 f_6 e_7 e_8^i$
$x_2^2 a_3 e_8^i$ $(i \geq 0)$	0	$x_2^2 a_3 e_8^{i+1}$	0	0	$x_1 x_2^2 f_6 e_7 e_8^i$
$e_7 e_8^i$ $(i \geq 0)$	0	$e_7 e_8^{i+1}$	$x_2^2 f_{11} e_8^i$	0	0
$x_2 e_7 e_8^i$ $(i \geq 0)$	0	$x_2 e_7 e_8^{i+1}$	0	0	0
$x_2^2 e_7 e_8^i$ $(i \geq 0)$	0	$x_2^2 e_7 e_8^{i+1}$	0	0	0
$f_6 e_8^i$ $(i \geq 0)$	$f_6 e_7 e_8^i$	$f_6 e_8^{i+1}$	0	0	0
$x_2 f_6 e_8^i$ $(i \geq 0)$	$x_2 f_6 e_7 e_8^i$	$x_2 f_6 e_8^{i+1}$	0	0	0
$x_2^2 f_6 e_8^i$ $(i \geq 0)$	$x_2^2 f_6 e_7 e_8^i$	$x_2^2 f_6 e_8^{i+1}$	0	0	0
f_9	0	$-x_2^2 f_6 e_7$	0	0	0
$f_9 a_2^i$ $(i \geq 1)$	0	0	0	0	0
$f_{12} e_8^i$ $(i \geq 0)$	0	$f_{12} e_8^{i+1}$	0	0	0
$x_2 f_{12} e_8^i$ $(i \geq 0)$	0	$x_2 f_{12} e_8^{i+1}$	0	0	0
$f_6 e_7 e_8^i$ $(i \geq 0)$	0	$f_6 e_7 e_8^{i+1}$	0	0	0
$x_2 f_6 e_7 e_8^i$ $(i \geq 0)$	0	$x_2 f_6 e_7 e_8^{i+1}$	0	0	0
$x_2^2 f_6 e_7 e_8^i$ $(i \geq 0)$	0	$x_2^2 f_6 e_7 e_8^{i+1}$	0	0	0
x_1	$x_1 e_7$	$x_1 e_8$	0	$a_2 f_8$	0
$x_1 e_8^i$ $(i \geq 1)$	$x_1 e_7 e_8^i$	$x_1 e_8^{i+1}$	0	$-x_1 x_2^2 f_6 e_7 e_8^{i-1}$	0
$x_1 x_2 e_8^i$ $(i \geq 0)$	$x_1 x_2 e_7 e_8^i$	$x_1 x_2 e_8^{i+1}$	0	0	0

	e_7	e_8	f_8	f_9	f_{11}
$x_1x_2^2e_8^i$ $(i \geq 0)$	$x_1x_2^2e_7e_8^i$	$x_1x_2^2e_8^{i+1}$	0	0	0
$x_1a_2^i$ $(i \geq 1)$	0	0	0	$a_2^{i+1}f_8$	0
$x_1a_3e_8^i$ $(i \geq 0)$	0	$x_1a_3e_8^{i+1}$	0	0	0
$x_1x_2a_3e_8^i$ $(i \geq 0)$	0	$x_1x_2a_3e_8^{i+1}$	0	0	0
$x_1e_7e_8^i$ $(i \geq 0)$	0	$x_1e_7e_8^{i+1}$	0	0	0
$x_1x_2e_7e_8^i$ $(i \geq 0)$	0	$x_1x_2e_7e_8^{i+1}$	0	0	0
$x_1x_2^2e_7e_8^i$ $(i \geq 0)$	0	$x_1x_2^2e_7e_8^{i+1}$	0	0	0
$x_1f_6e_8^i$ $(i \geq 0)$	$x_1f_6e_7e_8^i$	$x_1f_6e_8^{i+1}$	0	0	0
$x_1x_2f_6e_8^i$ $(i \geq 0)$	$x_1x_2f_6e_7e_8^i$	$x_1x_2f_6e_8^{i+1}$	0	0	0
$x_1x_2^2f_6e_8^i$ $(i \geq 0)$	$x_1x_2^2f_6e_7e_8^i$	$x_1x_2^2f_6e_8^{i+1}$	0	0	0
$f_8e_8^i$ $(i \geq 0)$	$x_2^2f_{11}e_8^i$	$f_8e_8^{i+1}$	0	0	0
a_2f_8	0	$-x_1x_2^2f_6e_8$	0	0	0
$a_2^if_8$ $(i \geq 2)$	0	0	0	0	0
$f_{11}e_8^i$ $(i \geq 0)$	0	$f_{11}e_8^{i+1}$	0	0	0
$x_2f_{11}e_8^i$ $(i \geq 0)$	0	$x_2f_{11}e_8^{i+1}$	0	0	0
$x_2^2f_{11}e_8^i$ $(i \geq 0)$	0	$x_2^2f_{11}e_8^{i+1}$	0	0	0
$x_1f_6e_7e_8^i$ $(i \geq 0)$	0	$x_1f_6e_7e_8^{i+1}$	0	0	0
$x_1x_2f_6e_7e_8^i$ $(i \geq 0)$	0	$x_1x_2f_6e_7e_8^{i+1}$	0	0	0
$x_1x_2^2f_6e_7e_8^i$ $(i \geq 0)$	0	$x_1x_2^2f_6e_7e_8^{i+1}$	0	0	0

This completes the proof of the lemma. □

Since all the differentials in the E_r-term for $r \geq 4$ vanish, it follows that $E_4 \cong \text{Cotor}_A(A, \mathbb{Z}/3)$ as an A-module.

Finally, it is necessary to solve the extension problem. It is easy to show that a_i for $i = 2, 3$, e_i for $i = 7, 8, 12$ and f_i for $i = 6, 7, 9, 11$ and 12 are represented by y_i for $i = 2, 3, 7, 8$ and 12 and z_i for $i = 6, 7, 9, 11$, and 12 respectively. The relations in $\text{Cotor}_A(A, \mathbb{Z}/3)$ corresponding to those in the E_∞-term are as follows:

the E_∞-term	$\text{Cotor}_A(A, \mathbb{Z}/3)$
x_2a_2	$x_2y_2 + x_1y_3$
a_2a_3	y_2y_3
x_2^3	x_2^3
$x_1x_2^2a_3$	$x_1x_2^2y_3$
a_2f_6	$y_2z_6 + x_1y_7$
x_1f_8	x_1z_8
a_2e_7	y_2y_7

9. PROOF OF THEOREM 1.6

$f_6 a_3$ $z_6 y_3 - x_2 y_7 - x_1 y_8$

$x_1 f_9 - a_2 f_8$ $x_1 z_9 - y_2 z_8$

$x_2 f_8$ $x_2 z_8$

$a_2 e_8 + a_3 e_7$ $y_2 y_8 + y_3 y_7$

$x_1 a_2 e_8$ $x_1 y_2 y_8$

$f_8 a_3 - x_1 x_2^2 f_6$ $z_8 y_3 - x_1 x_2^2 z_6$

$x_1 f_{11}$ $x_1 z_{11}$

$x_2 f_9$ $x_2 z_9$

$f_9 a_3$ $z_9 y_3 - x_1 x_2^2 y_7$

f_6^2 $z_6^2 + z_{12}$

$x_1 f_{12} - x_2 f_{11}$ $x_1 z_{12} - x_2 z_{11}$

$a_2 f_{11}$ $y_2 z_{11}$

$a_2 f_{12}$ $y_2 z_{12} - x_1 z_6 y_7$

$f_{11} a_3 + x_1 f_6 e_7$ $z_{11} y_3 + x_1 z_6 y_7$

$f_6 f_8$ $z_6 z_8$

$f_{12} a_3 + x_2 f_6 e_7$ $z_{12} y_3 + x_2 z_6 y_7 + x_1 z_6 y_8$

$f_6 f_9$ $z_6 z_9 + z_8 y_7$

$f_8 e_7 - x_2^2 f_{11}$ $z_8 y_7 - x_2^2 z_{11}$

$x_2^2 f_{12}$ $x_2^2 z_{12} - z_8 y_8$

$f_9 e_7$ $z_9 y_7$

f_8^2 z_8^2

$x_2^2 f_6 e_7 + f_9 e_8$ $x_2^2 z_6 y_7 + z_9 y_8$

$f_6 f_{11}$ $z_6 z_{11} + x_1 x_2^2 y_{12}$

$f_8 f_9$ $z_8 z_9$

$f_6 f_{12}$ $z_6 z_{12}$

$f_{11} e_7$ $z_{11} y_7 + x_1 x_2 y_3 y_{12}$

$f_{12} e_7$ $z_{12} y_7 + z_{11} y_8 + x_2^2 y_3 y_{12}$

$f_8 f_{11}$ $z_8 z_{11}$

$f_8 f_{12}$ $z_8 z_{12}$

$f_9 f_{11}$ $z_9 z_{11}$

$$f_9 f_{12} \qquad\qquad z_9 z_{12}$$
$$f_{11} f_{12} \qquad\qquad z_{11} z_{12} - x_1 x_2^2 z_6 y_{12}$$
$$f_{12}^2 \qquad\qquad z_{12}^2$$

Then we see that z_{12} is a decomposable element. It is easy to show that the relations $x_1 x_2^2 y_3$, $x_1 y_2 y_8$, $y_2 z_{12} - x_1 z_6 y_7$, $z_{12} y_3 + x_2 z_6 y_7 + x_1 z_6 y_8$, $z_8 z_{12}$, $z_9 z_{12}$, $z_{11} z_{12} - x_1 x_2^2 z_6 y_{12}$ and z_{12}^2 are generated by the other relations, and hence we can remove them from the set of the generators of the ideal. $\square$

Now we consider the Eilenberg-Moore spectral sequence

$$_C E_2 = \mathrm{Cotor}_{H^*(PU(3);\mathbb{Z}/3)}(H^*(PU(3);\mathbb{Z}/3), \mathbb{Z}/3) \Longrightarrow H^*(BLPU(3);\mathbb{Z}/3).$$

The diagram

$$\begin{array}{ccccc}
G & \longrightarrow & EG & \longrightarrow & BG \\
\downarrow & & \downarrow & & \downarrow {\scriptstyle Bs} \\
G & \longrightarrow & G & \longrightarrow & BLG \\
\downarrow & & \downarrow & & \downarrow {\scriptstyle B\pi} \\
G & \longrightarrow & EG & \longrightarrow & BG
\end{array}$$

induces the homomorphisms

$$\mathrm{id} = Bs^* \circ B\pi^* : {}_C\bar{E}_r \xrightarrow{B\pi^*} {}_C E_r \xrightarrow{Bs^*} {}_C\bar{E}_r,$$

where ${}_C\bar{E}_r$ is the r-th term of the Rothenberg-Steenrod spectral sequence

$$_C \bar{E}_2 = \mathrm{Cotor}_{H^*(G;\mathbb{Z}/3)}(\mathbb{Z}/3, \mathbb{Z}/3) \Longrightarrow H^*(BG;\mathbb{Z}/3).$$

Then it is easy to obtain the following lemma.

LEMMA 9.4. *Suppose that the Rothenberg-Steenrod spectral sequence $\{{}_C\bar{E}_r\}$ collapses. Then the elements in $\mathrm{Im}\, B\pi^* \subset {}_C E_2$ are permanent cycles. Moreover we have $\mathrm{Im}\, d_r \subset \mathrm{Ker}\, Bs^* \subset {}_C E_r$.*

According to Theorem 4.10 of [**23**], the Rothenberg-Steenrod spectral sequence

$$_C\bar{E}_2 = \mathrm{Cotor}_{H^*(PU(3);\mathbb{Z}/3)}(\mathbb{Z}/3, \mathbb{Z}/3) \Longrightarrow H^*(BPU(3);\mathbb{Z}/3)$$

collapses, so we can apply the lemma. Since $\mathrm{Ker}\, Bs^*$ is the ideal generated by x_1, x_2, z_6, z_8, z_9 and z_{11}, we can put the differential d_2 as follows:

$$d_2(z_6) = a_1 x_1 y_2^3, \quad d_2(z_8) = a_2 x_2 y_7 + a_3 x_1 y_8, \quad d_2(x_i) = d_2(z_9) = d_2(z_{11}) = 0,$$

where $a_i \in \mathbb{Z}/3$. The following equations imply that $a_i = 0$:

$$0 = d_2(-x_1 y_7) = d_2(y_2 z_6) = a_1 x_1 y_2^4,$$
$$0 = d_2(-x_2^2 z_6^2) = d_2(z_8 y_8) = a_2 x_2 y_7 y_8 + a_3 x_1 y_8^2.$$

Then we obtain $_CE_3 = {}_CE_2$. It is easy to see that $d_3 = 0$. Using the lemma, we can put the differential d_4 as follows:

$$d_4(z_8) = a_4 x_1 y_2^4, \quad d_4(x_i) = d_4(z_6) = d_4(z_9) = d_4(z_{11}) = 0,$$

where $a_4 \in \mathbb{Z}/3$. The following equation implies that $a_4 = 0$:

$$0 = d_4(x_1 z_9) = d_4(y_2 z_8) = a_4 x_1 y_2^5.$$

Then we obtain $_CE_5 = {}_CE_4$. It is easy to see that $_CE_\infty = {}_CE_5$.

10. Computation of a cotorsion product of $H^*(Spin(10); \mathbb{Z}/2)$ and the Hochschild homology of $H^*(BSpin(10); \mathbb{Z}/2)$

In this section, A and B denote the Hopf algebra $H^*(Spin(10); \mathbb{Z}/2)$ and the cohomology algebra

$$H^*(BSpin(10); \mathbb{Z}/2) = \mathbb{Z}/2[y_4, y_6, y_7, y_8, y_{10}, y_{32}]/(y_7 y_{10}),$$

respectively.

THEOREM 10.1. *As an $H^*(BSpin(10); \mathbb{Z}/2)$-algebra,*

$\mathrm{Cotor}_{H^*(Spin(10); \mathbb{Z}/2)}(H^*(Spin(10); \mathbb{Z}/2), \mathbb{Z}/2)$

$$\cong \mathbb{Z}/2[x_3, y_4, y_6, y_7, y_8, y_{10}, y_{32}] \otimes \Lambda(x_5, x_7, x_9, z_{30}, z_{31}, w_{31})/I,$$

in which $y_7 = \{\theta(x_3^2)\}$, $y_{i+1} = \{\theta x_i\}$ for $i = 3, 5, 7,$ and 9, $y_{32} = \{(\theta z_{15})^2\}$, $z_{30} = \{x_3^2 x_9 x_{15}\}$, $z_{31} = \{x_3^2 z_{15} \theta x_9 + x_3^2 x_9 \theta z_{15}\}$ and $w_{31} = \{x_9 x_{15} \theta x_6 + x_3^2 x_9 \theta z_{15}\}$, where $\{u\}$ denotes the element in the cotorsion product represented by a cocycle u in W'. Here I is the ideal generated by elements

$$x_3^4, \quad y_7 y_{10}, \quad w_{31} y_{10}, \quad z_{31} y_7, \quad x_3^2 z_{30}, \quad x_3^2 z_{31}, \quad x_9 y_7 + x_3^2 y_{10},$$

$$x_9 z_{30}, \quad x_9 w_{31}, \quad z_{30} z_{31}, \quad z_{30} w_{31}, \quad z_{31} w_{31}, \quad x_3^2 w_{31} + z_{30} y_7, \quad x_9 z_{31} + z_{30} y_{10},$$

and the bidegree of each element is given as follows: bideg $x_j = (0, j)$, bideg $y_{1+i} = (1, i)$ for $i + 1 = 4, 6, 7, 8, 10$, bideg $y_{32} = (2, 30)$, bideg $z_{30} = (0, 30)$, bideg $z_{31} = (1, 30)$, bideg $w_{31} = (1, 30)$.

PROOF. We denote θx_i, θx_3^2, θz_{15} by a_i, a_7, b_{16} respectively. We define a filtration of the complex $(W', d_{W'})$ by

$$F_i = \{x \otimes c \in A \otimes \overline{X} \mid |c| \geq i\}.$$

Consider the spectral sequence associated with the filtration converging to $\mathrm{Cotor}_A(A, \mathbb{Z}/2)$. The E_1-term is $A \otimes \overline{X}$ with the differential

$$d_1(x_i) = 0, \quad d_1(z_{15}) = 0, \quad d_1(a_i) = 0, \quad d_1(b_{16}) = a_7 a_{10}.$$

Then the E_2-term is isomorphic to $A \otimes \text{Cotor}_A(\mathbb{Z}/2, \mathbb{Z}/2)$ as an algebra. We have

$$\text{Cotor}_A(\mathbb{Z}/2, \mathbb{Z}/2) = \mathbb{Z}/2[a_4, a_7, a_6, a_8, a_{10}, c_{32}]/(a_7 a_{10}),$$

where c_{32} is represented by b_{16}^2. The first non-zero differential is given by

$$d_7(x_{15}) = x_9 a_7.$$

In order to calculate the E_8-term, we consider the following differential graded module:

$$(M, d) = (\Lambda(x_9, x_{15}) \otimes (\mathbb{Z}/2[a_{10}] \oplus \mathbb{Z}/2[a_7]\{a_7\}), d),$$

where $d(x_{15}) = x_9 a_7$.

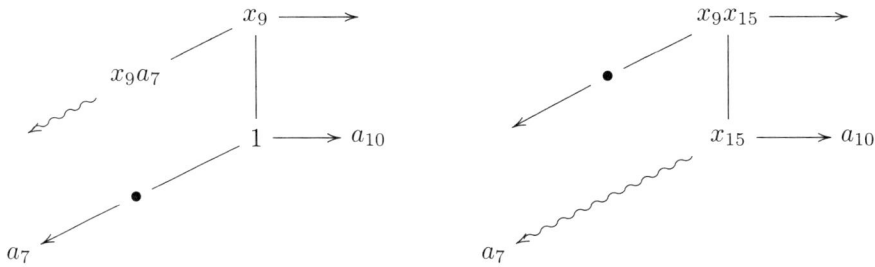

It is easy to calculate

$$H(M) = \mathbb{Z}/2[a_{10}]\{1, x_9, c_{24}, c_{25}\} \oplus \mathbb{Z}/2[a_7]\{a_7, a_7 c_{24}\},$$

where $c_{24} = x_9 x_{15}, c_{25} = x_{15} a_{10}$. We write

$$(E_7, d_7) = (\mathbb{Z}/2[x_3]/(x_3^4) \otimes \Lambda(x_5, x_7) \otimes \mathbb{Z}/2[a_4, a_6, a_8, c_{32}], 0) \otimes (M, d).$$

Thus we have

$$E_8 = \mathbb{Z}/2[x_3]/(x_3^4) \otimes \Lambda(x_5, x_7) \otimes \mathbb{Z}/2[a_4, a_6, a_8, c_{32}]$$
$$\otimes (\mathbb{Z}/2[a_{10}]\{1, x_9, c_{24}, c_{25}\} \oplus \mathbb{Z}/2[a_7]\{a_7, a_7 c_{24}\}).$$

LEMMA 10.2. *The algebra structure of the E_8-term is given by*

$$\mathbb{Z}/2[x_3]/(x_3^4) \otimes \Lambda(x_5, x_7, x_9) \otimes \mathbb{Z}/2[a_4, a_6, a_7, a_8, a_{10}, c_{32}] \otimes \Lambda(c_{24}, c_{25})/I,$$

where I is generated by

$$I = (x_9 a_7, a_7 a_{10}, c_{25} a_7, x_9 c_{24}, x_9 c_{25} + c_{24} a_{10}, c_{24} c_{25}).$$

PROOF. First we check that c_{24}^2, c_{25}^2 and generators of I are zero in the E_8-term. It is easy to see that $x_9 c_{24}$, $x_9 c_{25} + c_{24} a_{10}$, c_{24}^2, $c_{24} c_{25}$ and c_{25}^2 are zero in the chain level, that $a_7 a_{10}$ and $c_{25} a_7$ are zero in the E_2-term and that $x_9 a_7$ is in Im d_7. In order to check that there are no more relations, it is sufficient to show that, using the relations, the action of the product of the generators of the E_8-term are closed in the module structure. It is obvious that the action of the product of x_3, x_5, x_7, a_4, a_6, a_8 and c_{32} are closed. The following table is that of the other generators.

	x_9	a_7	a_{10}	c_{24}	c_{25}
a_{10}^i ($i \geq 1$)	$x_9 a_{10}^i$	0	a_{10}^{i+1}	$a_{10}^i c_{24}$	$a_{10}^i c_{25}$
$x_9 a_{10}^i$ ($i \geq 0$)	0	0	$x_9 a_{10}^{i+1}$	0	$a_{10}^{i+1} c_{24}$
c_{24}	0	$a_7 c_{24}$	$a_{10} c_{24}$	0	0
$a_{10}^i c_{24}$ ($i \geq 1$)	0	0	$a_{10}^{i+1} c_{24}$	0	0
$a_{10}^i c_{25}$ ($i \geq 0$)	$a_{10}^{i+1} c_{24}$	0	$a_{10}^{i+1} c_{25}$	0	0
a_7^i ($i \geq 1$)	0	a_7^{i+1}	0	$a_7^i c_{24}$	0
$a_7^i c_{24}$ ($i \geq 1$)	0	$a_7^{i+1} c_{24}$	0	0	0

This completes the proof of the lemma. □

The next non-zero differentials appear in the E_{10}-term and are given by

$$d_{10}(c_{24}) = x_3^2 x_9 a_{10}, \quad d_{10}(c_{25}) = x_3^2 a_{10}^2.$$

In order to calculate the E_{11}-term, we consider the following differential graded module:

$$(M, d) = (\Delta(x_3^2) \otimes (\mathbb{Z}/2[a_{10}]\{1, x_9, c_{24}, c_{25}\} \oplus \mathbb{Z}/2[a_7]\{a_7, a_7 c_{24}\}), d),$$

where $d(c_{24}) = x_3^2 x_9 a_{10}$ and $d(c_{25}) = x_3^2 a_{10}^2$.

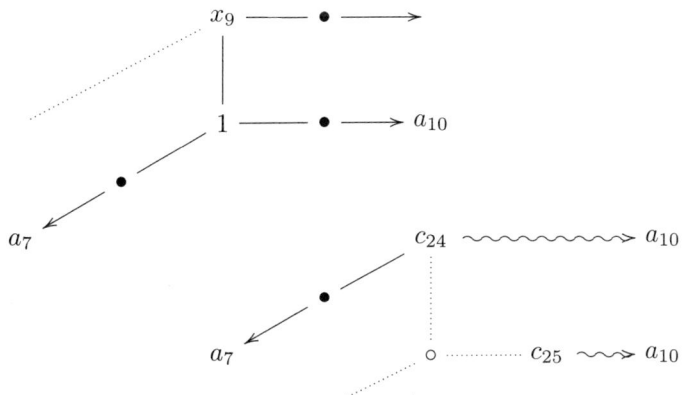

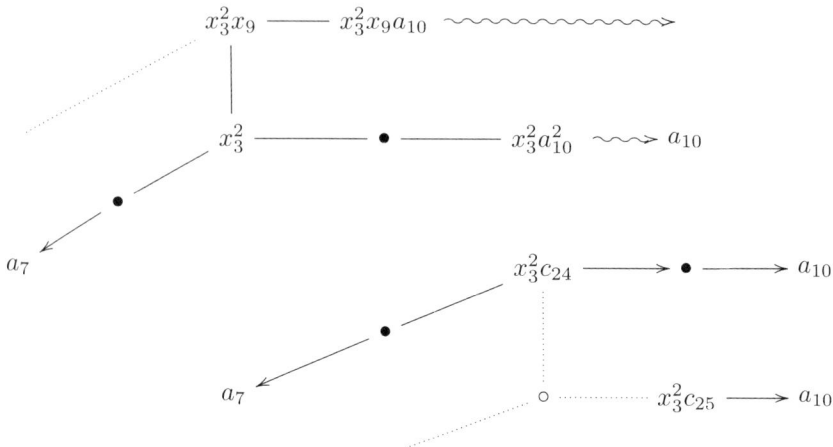

It is easy to calculate

$$H(M) = \mathbb{Z}/2[a_{10}]\{1, x_9, c_{30}, c_{31}\} \oplus \mathbb{Z}/2[a_7]\{a_7, x_3^2, e_{31}, c_{30}a_7\} \oplus \mathbb{Z}/2\{x_3^2 x_9, x_3^2 a_{10}\},$$

where $c_{30} = x_3^2 c_{24}, c_{31} = x_3^2 c_{25}, e_{31} = c_{24}a_7$. Since we have

$$(E_{10}, d_{10}) = (\Delta(x_3) \otimes \Lambda(x_5, x_7) \otimes \mathbb{Z}/2[a_4, a_6, a_8, c_{32}], 0) \otimes (M, d),$$

we obtain

$$E_{11} = \Lambda(x_5, x_7) \otimes \mathbb{Z}/2[a_4, a_6, a_8, c_{32}] \otimes \Delta(x_3)$$
$$\otimes (\mathbb{Z}/2[a_{10}]\{1, x_9, c_{30}, c_{31}\} \oplus \mathbb{Z}/2[a_7]\{a_7, x_3^2, e_{31}, c_{30}a_7\} \oplus \mathbb{Z}/2\{x_3^2 x_9, x_3^2 a_{10}\}).$$

LEMMA 10.3. *The algebra structure of the E_{11}-term is given by*

$$\mathbb{Z}/2[x_3] \otimes \Lambda(x_5, x_7, x_9) \otimes \mathbb{Z}/2[a_4, a_6, a_7, a_8, a_{10}, c_{32}] \otimes \Lambda(c_{30}, c_{31}, e_{31})/I,$$

where I is generated by

$$I = (x_3^4, x_9 a_7, a_7 a_{10}, x_3^2 x_9 a_{10}, x_3^2 a_{10}^2, x_3^2 c_{30}, x_3^2 c_{31}, x_3^2 e_{31} + c_{30} a_7, c_{31} a_7, x_9 c_{30},$$
$$x_9 c_{31} + c_{30} a_{10}, x_9 e_{31}, e_{31} a_{10}, c_{30} c_{31}, c_{30} e_{31}, c_{31} e_{31}).$$

PROOF. First we check that c_{30}^2, c_{31}^2, e_{31}^2 and generators of I are zero in the E_{11}-term. It is obvious that x_3^4, $x_3^2 c_{30}$, $x_3^2 c_{31}$, $x_3^2 e_{31} + c_{30} a_7$, $x_9 c_{30}$, $x_9 c_{31} + c_{30} a_{10}$, $x_9 e_{31}$, c_{30}^2, c_{31}^2, e_{31}^2, $c_{30} c_{31}$, $c_{30} e_{31}$ and $c_{31} e_{31}$ are zero in the chain level, that $a_7 a_{10}$, $c_{31} a_7$ and $e_{31} a_{10}$ are zero in the E_2-term, that $x_9 a_7$ is zero in the E_8-term and that $x_3^2 x_9 a_{10}$ and $x_3^2 a_{10}^2$ are in Im d_{10}. In order to check that there are no more relations, it is sufficient to show that, using the relations, the action of the product of the generators of the E_{11}-term are closed in the module structure. It is obvious that the action of the product of x_5, x_7, a_4, a_6, a_8 and c_{32} are closed. The following

10. A COTORSION PRODUCT AND THE HOCHSCHILD HOMOLOGY

table is that of the other generators.

	x_3	x_9	a_7	a_{10}	c_{30}	c_{31}	e_{31}
a_{10}^i $(i \geq 1)$	$x_3 a_{10}^i$	$x_9 a_{10}^i$	0	a_{10}^{i+1}	$a_{10}^i c_{30}$	$a_{10}^i c_{31}$	0
x_3	0	$x_3 x_9$	$x_3 a_7$	$x_3 a_{10}$	$x_3 c_{30}$	$x_3 c_{31}$	$x_3 e_{31}$
$x_3 a_{10}^i$ $(i \geq 1)$	0	$x_3 x_9 a_{10}^i$	0	$x_3 a_{10}^{i+1}$	$x_3 a_{10}^i c_{30}$	$x_3 a_{10}^i c_{31}$	0
$x_9 a_{10}^i$ $(i \geq 0)$	$x_3 x_9 a_{10}^i$	0	0	$x_9 a_{10}^{i+1}$	0	$a_{10}^{i+1} c_{30}$	0
$x_3 x_9 a_{10}^i$ $(i \geq 0)$	0	0	0	$x_3 x_9 a_{10}^{i+1}$	0	$x_3 a_{10}^{i+1} c_{30}$	0
c_{30}	$x_3 c_{30}$	0	$c_{30} a_7$	$a_{10} c_{30}$	0	0	0
$a_{10}^i c_{30}$ $(i \geq 1)$	$x_3 a_{10}^i c_{30}$	0	0	$a_{10}^{i+1} c_{30}$	0	0	0
$x_3 c_{30}$	0	0	$x_3 c_{30} a_7$	$x_3 a_{10} c_{30}$	0	0	0
$x_3 a_{10}^i c_{30}$ $(i \geq 1)$	0	0	0	$x_3 a_{10}^{i+1} c_{30}$	0	0	0
$a_{10}^i c_{31}$ $(i \geq 0)$	$x_3 a_{10}^i c_{31}$	$a_{10}^{i+1} c_{30}$	0	$a_{10}^{i+1} c_{31}$	0	0	0
$x_3 a_{10}^i c_{31}$ $(i \geq 0)$	0	$x_3 a_{10}^{i+1} c_{30}$	0	$x_3 a_{10}^{i+1} c_{31}$	0	0	0
a_7^i $(i \geq 1)$	$x_3 a_7^i$	0	a_7^{i+1}	0	$c_{30} a_7^i$	0	$e_{31} a_7^i$
$x_3 a_7^i$ $(i \geq 1)$	$x_3^2 a_7^i$	0	$x_3 a_7^{i+1}$	0	$x_3 c_{30} a_7^i$	0	$x_3 e_{31} a_7^i$
x_3^2	x_3^3	0	$x_3^2 a_7$	$x_3^2 a_{10}$	0	0	$c_{30} a_7$
$x_3^2 a_7^i$ $(i \geq 1)$	$x_3^3 a_7^i$	0	$x_3^2 a_7^{i+1}$	0	0	0	$c_{30} a_7^{i+1}$
x_3^3	0	0	$x_3^3 a_7$	$x_3^3 a_{10}$	0	0	$x_3 c_{30} a_7$
$x_3^3 a_7^i$ $(i \geq 1)$	0	0	$x_3^3 a_7^{i+1}$	0	0	0	$x_3 c_{30} a_7^{i+1}$
$e_{31} a_7^i$ $(i \geq 0)$	$x_3 e_{31} a_7^i$	0	$e_{31} a_7^{i+1}$	0	0	0	0
$x_3 e_{31} a_7^i$ $(i \geq 0)$	$c_{31} a_7^{i+1}$	0	$x_3 e_{31} a_7^{i+1}$	0	0	0	0
$c_{30} a_7^i$ $(i \geq 1)$	$x_3 c_{30} a_7^i$	0	$c_{30} a_7^{i+1}$	0	0	0	0
$x_3 c_{30} a_7^i$ $(i \geq 1)$	0	0	$x_3 c_{30} a_7^{i+1}$	0	0	0	0
$x_3^2 x_9$	$x_3^3 x_9$	0	0	0	0	0	0
$x_3^3 x_9$	0	0	0	0	0	0	0
$x_3^2 a_{10}$	$x_3^3 a_{10}$	0	0	0	0	0	0
$x_3^3 a_{10}$	0	0	0	0	0	0	0

This completes the proof of the lemma. $\square$

Since all the differentials in the E_r-term for $r \geq 11$ vanish, it follows that $E_{11} \cong \mathrm{Cotor}_A(\mathbb{Z}/2, \mathbb{Z}/2)$ as an A-module.

Finally, it is necessary to solve the extension problem. It is easy to show that a_i for $i = 4, 6, 7, 8$ and 10, c_{32}, c_{30}, c_{31} and e_{31} are represented by y_i for $i = 4, 6, 7, 8$ and 10, y_{32}, z_{30}, z_{31} and w_{31} respectively. The relations in $\mathrm{Cotor}_A(A, \mathbb{Z}/2)$ corresponding to those in the E_∞-term are as follows:

the E_∞-term	$\text{Cotor}_A(A, \mathbb{Z}/2)$
x_3^4	x_3^4
$x_9 a_7$	$x_9 y_7 + x_3^2 y_{10}$
$a_7 a_{10}$	$y_7 y_{10}$
$x_3^2 x_9 a_{10}$	$x_3^2 x_9 y_{10}$
$x_3^2 a_{10}^2$	$x_3^2 y_{10}^2$
$x_3^2 c_{30}$	$x_3^2 z_{30}$
$x_3^2 c_{31}$	$x_3^2 z_{31}$
$x_3^2 e_{31} + c_{30} a_7$	$x_3^2 w_{31} + z_{30} y_7$
$c_{31} a_7$	$z_{31} y_7$
$x_9 c_{30}$	$x_9 z_{30}$
$x_9 c_{31} + c_{30} a_{10}$	$x_9 z_{31} + z_{30} y_{10}$
$x_9 e_{31}$	$x_9 w_{31}$
$e_{31} a_{10}$	$w_{31} y_{10}$
c_{30}^2	z_{30}^2
$c_{30} c_{31}$	$z_{30} z_{31}$
$c_{30} e_{31}$	$z_{30} w_{31}$
c_{31}^2	z_{31}^2
$c_{31} e_{31}$	$z_{31} w_{31}$
e_{31}^2	w_{31}^2

The relations $x_3^2 x_9 y_{10}$ and $x_3^2 y_{10}^2$ are generated by $x_9 y_7 + x_3^2 y_{10}$ and $y_7 y_{10}$ respectively, and hence we can remove them from the set of the generators of the ideal. □

We turn to computing the Hochschild homology of B, as well as the torsion product $\text{Tor}_{B \otimes B}(B, B)$. The Koszul-Tate resolution of B as a $B \otimes B$-module is given by

$$\mathcal{F} \to B \otimes B \xrightarrow{m} B \to 0, \quad \mathcal{F} = B \otimes B \otimes \Lambda(\bar{w}_4, \bar{w}_6, \bar{w}_7, \bar{w}_8, \bar{w}_{10}, \bar{w}_{32}) \otimes \Gamma[\rho]$$

with differential

$$d(\gamma_i(\rho)) = (w_{10} \otimes 1 \otimes \bar{w}_7 + 1 \otimes w_7 \otimes \bar{w}_{10}) \gamma_{i-1}(\rho),$$

10. A COTORSION PRODUCT AND THE HOCHSCHILD HOMOLOGY   69

where bideg $\bar{w}_i = (-1, i)$ and bideg $\rho = (-2, 17)$. For more details of the construction of the resolution, we refer to [**42**](see also [**26**]).

Let $(\mathbf{C}(B), d)$ be the Hochschild complex of B with the shuffle product (see Section 8).

LEMMA 10.4. (see for example [**27**, Remark 3.3, Proposition 3.4]) *There exists a morphism of B-modules*

$$\Xi : (\mathbf{C}(B), d) \longrightarrow (B \otimes_{B \otimes B} \mathcal{F}, 1 \otimes d)$$

such that $\Xi(b) = b$ *for any* $b \in B$, $\Xi(1[w_i]) = \bar{w}_i$ *and* $\Xi(1[w_7|w_{10}]) = \rho$. *Moreover, the induced map*

$$H(\Xi) : HH(B) \longrightarrow \mathrm{Tor}_{B \otimes B}(B, B)$$

is an isomorphism of algebras.

The manner of computing a homology described in Section 7 is applicable to the Koszul-Tate complex $(B \otimes_{B \otimes B} \mathcal{F}, 1 \otimes d)$. In fact, Example 7.1 enables us to obtain the following theorem.

THEOREM 10.5. *As B-algebras,*

$$HH(B) \cong \mathrm{Tor}_{B \otimes B}(B, B)$$
$$\cong \mathbb{Z}/2[w_4, w_6, w_7, w_{10}, w_8, w_{32}]$$
$$\otimes \Lambda(\bar{w}_4, \bar{w}_6, \bar{w}_7, \bar{w}_8, \bar{w}_{10}, \bar{w}_{32}, (\bar{w}_7 \bar{w}_{10} \rho), (w_7 \bar{w}_{10} \rho))/I$$

for total degree < 45, *where* bideg $\bar{w}_i = (-1, i)$, bideg $(\bar{w}_7 \bar{w}_{10} \rho) = (-4, 34)$ *and* bideg $(w_7 \bar{w}_{10} \rho) = (-3, 34)$, *and I is generated by elements*

$w_7 w_{10}$, $\quad \bar{w}_7 w_{10} + w_7 \bar{w}_{10}$, $\quad (\bar{w}_7 \bar{w}_{10} \rho) w_7$, $\quad (\bar{w}_7 \bar{w}_{10} \rho) w_{10}$, $\quad (\bar{w}_7 \bar{w}_{10} \rho) \bar{w}_7$,

$(\bar{w}_7 \bar{w}_{10} \rho) \bar{w}_{10}$, $\quad (w_7 \bar{w}_{10} \rho) w_7$, $\quad (w_7 \bar{w}_{10} \rho) w_{10}$, $\quad (w_7 \bar{w}_{10} \rho) \bar{w}_7$, $\quad (w_7 \bar{w}_{10} \rho) \bar{w}_{10}$.

We are now ready to prove Lemma 1.8.

PROOF OF LEMMA 1.8. Recall that the spectral sequences $\{_{HH}E_r^{*,*}, d_r\}$ and $\{_C E_r^{*,*}, d_r\}$ converge to the same target $H^*(BLSpin(10); \mathbb{Z}/2)$. Comparing explicit forms of $_C E_2^{*,*}$ and $_{HH} E_2^{*,*}$ in Theorems 10.1 and 10.5 respectively, we see that, as a graded vector space,

$$\oplus_{i+j=*} {}_{HH}E_2^{i,j} \cong \oplus_{i+j=*} {}_C E_2^{i,j}$$

for $* < 45$. It follows from Theorem 10.1 that the bigraded algebra $_C E_2^{*,*}$ is generated by the elements with total degree ≤ 32. Thus we obtain Lemma 1.8.
□

11. Proof of Theorem 1.7

We shall prove Theorem 1.7 along the line described in Introduction.

LEMMA 11.1. *The elements of* $_{HH}E_2^{*,*}$ *with total degree* ≤ 29 *are permanent cycles and hence so are those of* $_C E_2^{*,*}$.

PROOF. Consider the spectral sequence $\{_{HH}E_r^{*,*}, d_r\}$. From Theorem 10.5, we see that the algebra generators of $_{HH}E_2^{*,*}$ with total degree ≤ 29 are in $_{HH}E_2^{-1,*}$. Therefore $\{_{HH}E_r^{*,*}, d_r\}$ collapses at the E_2-term for total degree ≤ 29. The result follows from Lemma 1.8. □

LEMMA 11.2. *The elements of* $_C E_2^{*,*}$ *with total degree* 30 *are permanent cycles and hence so are those of* $_{HH}E_2^{*,*}$.

PROOF. We consider the cobar type EMSS $\{_C E_r^{*,*}, d_r\}$. From Theorem 10.1, we have just one indecomposable element of total degree 30 in the E_2-term, which is represented by z_{30} in $_C E_2^{0,30}$. Therefore in order to prove the lemma, it suffices to show that $d_r(z_{30}) = 0$ for any $r \geq 2$. Let g be the map from the the twisted tensor product of $H^*(E_6; \mathbb{Z}/2)$ to that of $H^*(Spin(10); \mathbb{Z}/2)$ induced by the inclusion $Spin(10) \hookrightarrow E_6$. Observe that the map g induces a morphism $\{g_r\}$ of spectral sequences from the cobar type EMSS converging to $H^*(BLE_6; \mathbb{Z}/2)$ to $\{_C E_r^{*,*}, d_r\}$. In particular, the map g_2 is regarded as the morphism of algebras

$$h = H(1 \square g) : \text{Cotor}_{H^*(E_6; \mathbb{Z}/2)}(H^*(E_6; \mathbb{Z}/2), \mathbb{Z}/2)$$
$$\longrightarrow \text{Cotor}_{H^*(Spin(10); \mathbb{Z}/2)}(H^*(Spin(10); \mathbb{Z}/2), \mathbb{Z}/2).$$

Since the images of the elements x_{17}, x_{23}, a_{18} and b_{24} by the map g are zero, and since $g(x_i) = x_i$ for $i = 3, 5, 9, 15$, $g(a_j) = y_j$ for $j = 4, 7, 6, 10$ and $g(b_{16}) = b_{16}$, it follows that, for any $\alpha \in \text{Cotor}^{2,29}_{H^*(E_6;\mathbb{Z}/2)}(H^*(E_6;\mathbb{Z}/2), \mathbb{Z}/2)$, the image $h(\alpha)$ is a linear combination of the elements whose terms do not contain the elements x_7 and y_8 as a factor. Since there is an element $z' \in \text{Cotor}^{0,30}_{H^*(E_6;\mathbb{Z}/2)}(H^*(E_6;\mathbb{Z}/2), \mathbb{Z}/2)$ such that $g_2(z') = z_{30}$, it follows that $d_2(z_{30})$ is in the image of g_2. The Steenrod operations Sq^i in the EMSS [35] for $i = 1, 2, 4$ and 8 and the operations of the multiplication by the elements x_3^2 and x_9 vanish on the element z_{30}. Therefore so does such an action on the element $d_r(z_{30})$ for $r \geq 2$. By direct calculation, we can determine such an action on $_C E_2^{i,j}$ ($i + j = 31$), which is inherited by $_C E_r^{i,j}$ for any r. For the result, see Tables (A.1) in Appendix. If $d_2(z_{30}) \neq 0$, then, from Table (A.1), $d_2(z_{30})$ is written $k_1 x_3^2 a_7 a_8 a_{10} + k_2 x_3^2 x_9 a_8^2$, where $k_i \in \mathbb{Z}/2$ and $k_1 k_2 \neq 0$. We can conclude that $d_2(z_{30})$ is not in the image of h, which is a contradiction. From Tables (A,1) and the fact that the action mentioned above is trivial on $d_r(z_{30})$, it

11. PROOF OF THEOREM 1.7

follows that $d_r(z_{30}) = 0$ for any r. The latter half of the assertion follows from Lemma 1.8. □

In order to consider the differential on $_{HH}E_r^{*,*}$ with total degree 31, we make use of the TV-model for $BSpin(10)$.

We here review briefly general properties of TV-models. Let (TV, d) be the (minimal) TV-model for a simply connected space X, which exists uniquely up to homotopy. The result [**10**, Proposition A.8] states that the vector space V is isomorphic to $s\bar{H}^*(\Omega X; \mathbb{Z}/p)$ and the quadratic part of the differential d of TV coincides with the coproduct on $H^*(\Omega X; \mathbb{Z}/p)$. More precisely, if $\bar{\phi}(v) = \sum a_i' \otimes a_i''$, then the quadratic part of $d(sv)$, namely, the restriction on T^2V of the differential, is $\sum sa_i' sa_i''$.

From now on, we direct our attention to the TV-model (TV, d) for $BSpin(10)$ over $\mathbb{Z}/2$. Observe that

$$V \cong s\overline{H}^*(\Omega BSpin(10); \mathbb{Z}/2) \cong s\overline{H}^*(Spin(10); \mathbb{Z}/2).$$

As is recalled in the previous section,

$$H^*(Spin(10); \mathbb{Z}/2) \cong \mathbb{Z}/2[x_3]/(x_3^4) \otimes \Lambda(x_5, x_7, x_9, x_{15}),$$

where the elements x_i are primitive and $\bar{\phi}(z_{15}) = x_3^2 \otimes x_9$. We express the elements sx_3, sx_5, $s(x_3^2)$, sx_7 and sx_9 in V as w_4, w_6, w_7, w_8 and w_{10}, respectively. Since x_i ($i = 3, 5, 7, 9$) and x_3^2 are primitive, we can conclude that the quadratic part of dw_j are zero. The least degree of the elements in $T^3V = V \cdot V \cdot V$ is 12. Hence we see that the indecomposable elements w_j are cycles in TV. Since

$$H^*(BSpin(10); \mathbb{Z}/2) \cong \mathbb{Z}/2[w_4, w_6, w_7, w_8, w_{10}, w_{32}]/(w_7 w_{10})$$

as an algebra, we may assume that the elements w_j in TV represent the indecomposable elements w_j ($j = 4, 6, 7, 8, 10$) in $H^*(BSpin(10); \mathbb{Z}/2)$, respectively. Moreover, we see that there exist indecomposable elements ε and δ in TV such that $d\varepsilon = w_7 w_{10}$ and $d\delta = w_{10} w_7$. Obviously the elements $w_{10}\varepsilon + \delta w_{10}$ and $\varepsilon w_7 + w_7 \delta$ are cycles. Therefore there exist polynomials ξ_1 and ξ_2 consisting of the elements w_j and indecomposable elements γ and γ' in TV such that

$$d\gamma = w_{10}\varepsilon + \delta w_{10} + \xi_1,$$
$$d\gamma' = \varepsilon w_7 + w_7 \delta + \xi_2.$$

Since $d(w_7\gamma + \gamma' w_{10} + \varepsilon^2) = w_7 \xi_1 + \xi_2 w_{10}$, it follows that $w_7 \xi_1 + \xi_2 w_{10} = 0$ in $H^*(TV) \cong H^*(BSpin(10); \mathbb{Z}/2)$. Thus it is easily seen that ξ_1 and ξ_2 are elements in the ideals $(w_{10})_{H^*(TV)}$ and $(w_7)_{H^*(TV)}$, respectively. This fact allows us to write

$\xi_1 = w_{10}a + dv$ and $\xi_2 = bw_7 + dv'$ for some v and v' in TV. There is no loss in generality in supposing that

$$d\gamma = w_{10}\varepsilon + \delta w_{10} + w_{10}a,$$
$$d\gamma' = \varepsilon w_7 + w_7\delta + bw_7.$$

Moreover, we can choose the indecomposable elements γ, γ' and an appropriate element ξ so that

$$d\gamma = w_{10}\varepsilon + \delta w_{10},$$
$$d\gamma' = \varepsilon w_7 + w_7\delta + \xi w_7$$

and ξ is a polynomial of the elements w_4, w_6, w_7 and w_8. For dimensional reasons, we can see that ξ is a polynomial of w_4, w_6 and w_8, since ξ is of degree 16.

LEMMA 11.3. *The indecomposable elements $\bar{w}_{32}$ and $w_7\bar{w}_{10}\rho$ with degree 31 in*

$$_{HH}E_2^{*,*} \cong HH(H^*(BSpin(10); \mathbb{Z}/2)) \cong \mathrm{Tor}_{B\otimes B}(B, B)$$

are permanent cycles.

PROOF. Since the bidegree of $\bar{w}_{32}$ is $(-1, 32)$, it is immediate to see that $\bar{w}_{32}$ is a permanent cycle, as the spectral sequence under consideration is of the second quadrant. In the E_2-term of $\{_{HH}E_r^{*,*}, d_r\}$, namely, in the Hochschild homology $HH(H^*(BSpin(10); \mathbb{Z}/2))$, the element $w_7\bar{w}_{10}\rho$ is written

$$w_7 \cdot 1[w_{10}] \cdot 1[w_7|w_{10}] = w_7[w_{10}|w_7|w_{10}]$$

in terms of the Hochschild complex under the isomorphism in Lemma 10.4. Here the *dot* $\cdot$ stands for the shuffle product. Thus, in order to deduce the result, it suffices to show that $w_7[w_{10}|w_7|w_{10}]$ is a permanent cycle. To see this, we consider the Hochschild spectral sequence converging to $HH(TV, d)$ instead of $\{_{HH}E_r^{*,*}, d_r\}$. We choose elements

$$\alpha_0 = w_7[w_{10}|w_7|w_{10}],$$
$$\alpha_1 = \varepsilon[w_7|w_{10}] + w_7[\delta|w_{10}] + w_7[w_{10}|\varepsilon] + \delta[w_{10}|w_7],$$
$$\alpha_2 = \varepsilon[\varepsilon] + \gamma[w_7] + \delta[\delta] + w_7[\gamma],$$

in the Hochschild complex $(\mathbf{C}(TV), b)$. By abuse of notation, a cycle in TV and its cohomology class in $H(TV)$ are denoted by the same letter. By direct calculation, we see that $d_1\alpha_1 = d_2\alpha_2$, $d_1\alpha_2 = d_2\alpha_3$ and $d_1\alpha_3 = 0$. The result follows from [25, Lemma 2.1]. □

PROOF OF THEOREM 1.7. The element w_{32} in $_{HH}E_2^{0,32}$ is a permanent cycle. Therefore, by combining Lemmas 11.1, 11.2, 11.3 and 1.8, we see that the cobar type EMSS $\{_C E_r^{*,*}, d_r\}$ collapses at the E_2-term. In order to clarify the $H^*(BSpin(10); \mathbb{Z}/2)$-module structure of $H^*(BLSpin(10); \mathbb{Z}/2)$, we first consider the cobar type EMSS $\{\widetilde{E}_r^{*,*}, \widetilde{d}_r\}$ converging to $H^*(BSpin(10); \mathbb{Z}/2)$. Put $G = Spin(10)$ for simplicity. The trivial map $G \to \{e\}$ induces a morphism of spectral sequences

$$\{f_r\} : \{\widetilde{E}_r^{*,*}, \widetilde{d}_r\} \to \{_C E_r^{*,*}, d_r\}.$$

Let

$$f : G \times_{\mathrm{ad}} EG \to \{e\} \times_{\mathrm{ad}} EG = BG$$

be the map induced from the trivial map. The naturality of the spectral sequence implies that the induced morphism f_∞ coincides up to isomorphism with the map

$$E_0 H^*(f) : E_0 H^*(BG; \mathbb{Z}/2) \to E_0 H^*(BLG; \mathbb{Z}/2)$$

between the E_∞-terms and associated bigraded algebras. Observe that $H^*(f)$ gives the $H^*(G; \mathbb{Z}/2)$-module structure of $H^*(BLG; \mathbb{Z}/2)$. By using the twisted tensor product in Proposition 3.2, we have

$$\widetilde{E}_2^{*,*} \cong \mathbb{Z}/2[w_4, w_6, w_7, w_8, w_{10}, w_{32}]/(w_7 w_{10})$$

as a bigraded algebra. Moreover, we see that $f_2(w_i) = y_i$. The algebra structure $H^*(BG; \mathbb{Z}/2)$, which is known to us, allows us to conclude that the EMSS $\{\widetilde{E}_r^{*,*}, \widetilde{d}_r\}$ collapses at the E_2-term, and hence

$$H^*(BG; \mathbb{Z}/2) \cong \mathbb{Z}/2[w_4, w_6, w_7, w_8, w_{10}, w_{32}]/(w_7 w_{10} + \rho)$$

as an algebra, where ρ is an appropriate decomposable element. We can choose the elements w_7 and w_{10} so that $w_7 w_{10} = 0$ in $H^*(BG; \mathbb{Z}/2)$. Consequently, a suitable choice of representative elements of y_7 and y_{10} of $H^*(BLG; \mathbb{Z}/2)$ enables us to obtain the explicit $H^*(BG; \mathbb{Z}/2)$-module structure of $H^*(BG; \mathbb{Z}/2)$ stated in Theorem 1.7. □

12. Proofs of Proposition 1.9 and Theorem 1.10

By considering the differential d_2 on the element $\bar{w}_7 \bar{w}_{10} \rho$ in terms of the TV-model, we deduce the following proposition, which leads us to a proof of Theorem 1.10.

PROPOSITION 12.1. *Let TV be the TV-model, which is constructed in the previous section, for $BSpin(10)$ over $\mathbb{Z}/2$. We can choose an indecomposable element γ' in TV so that $d\gamma' = \varepsilon w_7 + w_7 \delta$.*

PROOF. As in the proof of Lemma 11.3, we can take an element

$$\beta_0 = [w_7] \cdot [w_{10}] \cdot [w_7|w_{10}] = [w_7|w_{10}|w_7|w_{10}] + [w_{10}|w_7|w_{10}|w_7]$$

in the Hochschild complex $\mathbf{C}(H(TV))$ which corresponds to the element $\bar{w}_7 \bar{w}_{10} \rho$ in the E_2-term of the HSS. We can write

$$d\gamma' = \varepsilon w_7 + w_7 \delta + P(w_4, w_6, w_8) w_7,$$

in which $P(w_4, w_6, w_8)$ is a polynomial consisting of w_4, w_6 and w_8 (see the result before Lemma 11.3). Define elements β_1 and β_2 in the Hochschild complex $\mathbf{C}(TV)$ as follows:

$$\beta_1 = [\delta|w_{10}|w_7] + [w_{10}|\varepsilon|w_7] + [w_{10}|w_7|\delta] + [\varepsilon|w_7|w_{10}] + [w_7|\delta|w_{10}] + [w_7|w_{10}|\varepsilon],$$
$$\beta_2 = [\gamma|w_{10}] + [\delta|\delta] + [w_{10}|\gamma'] + [\gamma'|w_{10}] + [\varepsilon|\varepsilon] + [w_7|\gamma].$$

By direct calculation, we see that $b_1 \beta_0 = b_0 \beta_1$ and

$$b_1 \beta_1 = b_0 \beta_2 + [w_{10}|P(w_4, w_6, w_8)w_7] + [P(w_4, w_6, w_8)w_7|w_{10}].$$

Therefore, we have

$$d_2(\bar{w}_7 \bar{w}_{10} \rho) = [w_{10}|P(w_4, w_6, w_8)w_7] + [P(w_4, w_6, w_8)w_7|w_{10}]$$
$$= [P(w_4, w_6, w_8)w_7] \cdot [w_{10}]$$

in the E_2-term of the spectral sequence $\{_{HH}E_r^{*,*}, d_r\}$. Since

$$(b_0 + b_1)[c_1|c_2] = c_1[c_2] + [c_1 c_2] + c_2[c_1]$$

for cycles c_1 and c_2, it follows that, in the E_2-term,

$$[P(w_4, w_6, w_8)w_7] = P(w_4, w_6, w_8)[w_7] + Q_1(w_4, w_6, w_8)w_7[w_4]$$
$$+ Q_2(w_4, w_6, w_8)w_7[w_6] + Q_3(w_4, w_6, w_8)w_7[w_8],$$

where $Q_i(w_4, w_6, w_8)$ is a polynomial of the elements w_4, w_6 and w_8. Thus we have

$$d_2(\bar{w}_7 \bar{w}_{10} \rho) = P(w_4, w_6, w_8)\bar{w}_7 \bar{w}_{10} + Q_1(w_4, w_6, w_8)w_7 \bar{w}_4 \bar{w}_{10}$$
$$+ Q_2(w_4, w_6, w_8)w_7 \bar{w}_6 \bar{w}_{10} + Q_3(w_4, w_6, w_8)w_7 \bar{w}_8 \bar{w}_{10}.$$

Suppose that $P(w_4, w_6, w_8) \neq 0$ in $H^*(TV) = H^*(BSpin(10); \mathbb{Z}/2)$. The algebra structure of the E_2-term of $\{_{HH}E_r, d_r\}$ (see Theorem 10.1) enables us to conclude that $d_2(\bar{w}_7 \bar{w}_{10} \rho) \neq 0$, which contradicts the fact that $\{_{HH}E_r, d_r\}$ collapses at the E_2-term. This completes the proof. □

Before proving Proposition 1.9 and Theorem 1.10, we recall a Hochschild homological description of the map $\lambda = \int_{S^1} \circ \alpha_{\mathbb{T}}$.

12. PROOFS OF PROPOSITION 1.9 AND THEOREM 1.10

THEOREM 12.2. [**18**, Theorem 4.1] *For any simply connected space X, there exists a commutative diagram*

$$\begin{array}{ccc} HH_*(C^*(X;\mathbb{Z}/p)) & \xrightarrow[\cong]{J} & H^*(X^{S^1};\mathbb{Z}/p) \\ B\downarrow & & \downarrow \lambda \\ HH_{*-1}(C^*(X;\mathbb{Z}/p)) & \xrightarrow[\cong]{J} & H^{*-1}(X^{S^1};\mathbb{Z}/p), \end{array}$$

where J is the isomorphism due to Jones and B is the Connes B-map.

For the definition of the Connes B-map, we refer the reader to Appendix.

PROOF OF PROPOSITION 1.9. Let A be the cochain complex $C^*(X;\mathbb{Z}/p)$. Let $F^* = \{F^{-i}HH(A); i \leq 0\}$ denote the filtration defined in the Hochschild spectral sequence converging to $HH_*(A)$; that is,

$$F^{-i}HH_*(A) = \mathrm{Im}\{j_* : H(\sum_{k \leq i} A \otimes \bar{A}^{\otimes k}, b_0 + b_1) \to HH_*(A)\},$$

where j is the natural inclusion of the subcomplex $\sum_{k \leq i} A \otimes \bar{A}^{\otimes k}$ into the Hochschild complex. From the definition of the Connes B-map, it follows that the map B decreases the filtration degree by 1. Define a map $\varphi_L : BLG \to (BG)^{S^1}$ by

$$\varphi_L(\gamma_1 \oplus \cdots \oplus \gamma_l)(t) = \gamma_1(t) \oplus \cdots \oplus \gamma_l(t).$$

It is immediate to see that φ_L is a $\mathbb{T}$-equivariant map. Moreover, it is known by [**28**] that φ_L is a homotopy equivalence. Combining these facts with Theorem 12.2, we have the result. □

PROOF OF THEOREM 1.10. Throughout this proof, we identify the HSS $\{_{HH}E_r^{*,*}, d_r\}$ converging to $H^*(BLSpin(10);\mathbb{Z}/2)$ with that converging to $HH_*(TV)$, where TV is the TV-model for $BSpin(10)$ stated in Proposition 12.1. As is seen in the previous sections, the E_∞-term of the cobar type EMSS is generated by the elements with total degree ≤ 32 and hence so is $_{HH}E_\infty^{*,*}$. Therefore representative elements of w_i, $\bar{w}_i$ ($i = 4, 6, 7, 8$ and 32), $\bar{w}_7\bar{w}_{10}\rho$ and $w_7\bar{w}_{10}\rho$ generate the cohomology algebra $H^*(LBSpin(10);\mathbb{Z}/2)$. Under the notation of Section 5, we can choose the elements w_i, $[w_i]$ and $\xi_{31} = \alpha_0 + \alpha_2 + \alpha_3$ as representatives of w_i, $\bar{w}_i$ and $w_7\bar{w}_{10}\rho$, respectively. An easy calculation gives formulae $B(w_i) = [w_i]$, $B[w_i] = 0$ and $B(\xi_{31}) = 0$. In the Hochschild complex $\mathbf{C}(TV)$ of TV, we have

$$(b_0 + b_1)(\beta_0 + \beta_1 + \beta_2) = [\gamma w_7 + \delta^2 + w_{10}\gamma'] + [w_7\gamma + \varepsilon^2 + \gamma'w_{10}].$$

Therefore it follows that

$$d_3(\bar{w}_7\bar{w}_{10}\rho) = [\gamma w_7 + \delta^2 + w_{10}\gamma'] + [w_7\gamma + \varepsilon^2 + \gamma'w_{10}]$$

in $_{HH}E_3^{*,*}$. Since $\{_{HH}E_r^{*,*}, d_r\}$ collapses at the E_2-term, we can choose a cycle ι_1 of the differential b_0 in $TV \otimes (\overline{TV})^{\otimes 2}$ and an element ι_2 in $TV \otimes (\overline{TV})$ so that

$$(b_0 + b_1)\iota_1 = (\gamma w_7 + \delta^2 + w_{10}\gamma') + (w_7\gamma + \varepsilon^2 + \gamma' w_{10}) + b_0\iota_2.$$

Thus we see that

$$(b_0 + b_1)(\beta_0 + \beta_1 + \beta_2 + \iota_1 + \iota_2) = b_1\iota_2.$$

Since d_4 is also trivial, we can take an element ι_3 in TV so that $b_1\iota_2 = b_0\iota_3$. It turns out that the element

$$\xi_{30} = \beta_0 + \beta_1 + \beta_2 + \iota_1 + \iota_2 + \iota_3$$

is a cycle in $\mathbf{C}(TV)$ and the element ξ_{30} in $F^{-4}HH_*(TV)$ represents the generator $\bar{w}_7\bar{w}_{10}\rho$ of $_{HH}E_\infty^{*,*}$. The definitions of the elements β_0, β_1 and β_2 allow us to deduce that $B(\beta_0 + \beta_1 + \beta_2) = 0$. Thus $B(\xi_{30})$ is in the filtration $F^{-3}HH_*(TV)$. The result follows from this fact and Theorem 12.2. □

Thanks to Proposition 12.1, we can represent the only indecomposable element in $H^*(BSpin(10); \mathbb{Z}/2)$ of degree 31 in terms of the TV-model for $BSpin(10)$.

PROPOSITION 12.3. *In the cohomology $H(TV) = H^*(BSpin(10); \mathbb{Z}/p)$, the elements $\{\gamma w_7 + \delta^2 + w_{10}\gamma'\}$ and $\{w_7\gamma + \varepsilon^2 + \gamma' w_{10}\}$ coincide with the only indecomposable element in $H^{32}(BSpin(10); \mathbb{Z}/2)$ modulo decomposable elements.*

PROOF. Let us consider the spectral sequence, which is described in Theorem 8.2, converging to $H^*(\Omega BSpin(10); \mathbb{Z}/2) \cong H^*(Spin(10); \mathbb{Z}/p)$. By using the Koszul-Tate resolution, and by applying the same argument as in the proof of [**26**, Lemma 1.5], we construct an isomorphism of algebras

$$\Psi : {}_\Omega E_2^{*,*} \cong H(BH^*(BSpin(10); \mathbb{Z}/2))$$
$$\xrightarrow{\cong} \Lambda(s^{-1}w_4, s^{-1}w_6, s^{-1}w_7, s^{-1}w_8, s^{-1}w_{10}, s^{-1}w_{32}) \otimes \Gamma[\rho]$$

such that $\Psi([w_i]) = s^{-1}w_i$ and

$$\Psi([w_7|w_{10}|w_7|w_{10}|\cdots|w_7|w_{10}]) = \gamma_l(\rho),$$

where $w_7|w_{10}$ appears 21 times. Since there is no indecomposable element of degree 30 in $H^*(Spin(10); \mathbb{Z}/2)$, it follows that $d_3(\gamma(\rho)) = s^{-1}w_{32}$, for dimensional reasons. This fact implies that

$$d_3([w_7|w_{10}|w_7|w_{10}]) = [w_{32}]$$

in terms of the bar complex. By the definition of the shuffle product on the bar complex, we see that

$$[w_7] \cdot [w_{10}] \cdot [w_7|w_{10}] = [w_7|w_{10}|w_7|w_{10}],$$

in the E_2-term. Since $d_3(\gamma(\rho) + s^{-1}w_7 s^{-1} w_{10}\rho) = s^{-1}w_{32}$, it follows that

$$d_3([w_{10}|w_7|w_{10}|w_7]) = [w_{32}].$$

Take the TV-model

$$(TV, d) \xrightarrow{\simeq} (C^*(BSpin(10), \mathbb{Z}/2), d)$$

which satisfies the condition in Proposition 12.1. We then identify the spectral sequence $\{_\Omega E_r, d_r\}$ with the algebraic spectral sequence $\{_\Omega \widetilde{E}_r, \widetilde{d}_r\}$ converging to $H^*(BTV, d_{BTV})$ with $_\Omega \widetilde{E}_2 \cong H(BH^*(TV))$. We choose elements

$$\alpha'_0 = [w_7|w_{10}|w_7|w_{10}],$$
$$\alpha'_1 = [\varepsilon|w_7|w_{10}] + [w_7|\delta|w_{10}] + [w_7|w_{10}|\varepsilon],$$
$$\alpha'_2 = [\gamma'|w_{10}] + [\varepsilon|\varepsilon] + [w_7|\gamma]$$

of the bar complex $B(TV, d)$. It is easy to verify that

$$d(\alpha'_0 + \alpha'_1 + \alpha'_2) = [\gamma' w_{10} + \varepsilon^2 + w_7\gamma].$$

This fact enables us to deduce that

$$\widetilde{d}_3([w_7|w_{10}|w_7|w_{10}]) = [\gamma' w_{10} + \varepsilon^2 + w_7\gamma]$$

in the E_3-term $_\Omega \widetilde{E}_3^{*,*}(\cong {}_\Omega \widetilde{E}_2^{*,*})$. Thus we have the result. The same argument as above works well to show that $\gamma w_7 + \delta^2 + w_{10}\gamma' \equiv w_{32}$ modulo decomposable elements in $H^*(TV) \cong H^*(BSpin(10); \mathbb{Z}/2)$. □

13. Appendix

In this section, we give tables concerning the action on the cobar type EMSS $_C E_2^{*,*}$ for total degree 31 mentioned in the proof of Lemma 11.2, namely, the action of the Steenrod operations Sq^i on the EMSS ($i = 1, 2, 4$ and 8) and the operations of the multiplication by the elements x_3^2 and x_9. The representatives of θx_i ($i = 3, 5, 7$ and 9) and $\{\theta(x_3^2)\}$ on the E_2-term $_C E_2^{*,*}$ of the cobar type EMSS will be denoted by a_{i+1} and a_7, respectively. Let A be the Hopf algebra $H^*(Spin(10); \mathbb{Z}/2)$. In the cobar type EMSS, the calculation of the action of the Steenrod operations is performed after representing elements as those in the homology of the cobar complex $D_A^*(\mathbb{Z}/2)$. Since the Steenrod operations on the cobar complex $D_A^*(\mathbb{Z}/2)$ satisfy the Cartan formula (see [**35**]), we determine explicitly the action. More precisely, if an element

$$\sum_{i_0,\ldots,i_l} a_{i_0} \otimes a_{i_1} \otimes \cdots \otimes a_{i_l}$$

in $D_A^*(\mathbb{Z}/2)$ represents an element $\alpha \in \mathrm{Cotor}_A(A, \mathbb{Z}/p)$. Then we have

$$Sq^k \alpha = \sum_{k_0+\cdots+k_l=k} \sum_{i_0,\ldots,i_l} Sq^{k_0} a_{i_0} \otimes Sq^{k_0} a_{i_1} a_{i_1} \otimes \cdots \otimes Sq^{k_l} a_{i_l}.$$

The isomorphism $\widetilde{\pi}$ defined in Remark 5.13 maps the element α to

$$\sum_{i_0,\ldots,i_l} a_{i_0} \theta a_{i_1} \cdots \theta a_{i_l}.$$

In consequence, for example, we have

$$Sq^4(x_3^3 x_5 x_9 \theta x_3 \theta x_3) = x_3^3 x_5 x_9 \theta(Sq^2 x_3) \theta(Sq^2 x_3) = x_3^3 x_5 x_9 \theta x_5 \theta x_5.$$

Tables (A.1).

$\mathrm{Cotor}^{2,29}$	x_3^2	x_9	Sq^1	Sq^2	Sq^4	Sq^8
$x_3^3 x_5 x_7 a_4 a_6$	0	$x_3^3 x_5 x_7 x_9 a_4 a_6$				
$x_3^3 x_5 x_9 a_4^2$	0	0	0	0	$x_3^3 x_5 x_9 a_6^2$	
$x_3^3 x_7 a_7 a_8$	0	0	0	$x_3^2 x_5 x_7 a_7 a_8$		
$x_3^2 x_5 x_7 a_6 a_7$	0	0	$x_3^2 x_5 x_7 a_7^2$			
$x_3^2 x_7 a_8 a_{10}$	0	0	0	0	0	
$x_3^2 x_9 a_8^2$	0	0	0	0	0	
$x_3 x_5 x_7 a_6 a_{10}$	$x_3^3 x_5 x_7 a_6 a_{10}$					
$x_3 x_5 x_7 a_8^2$	$x_3^3 x_5 x_7 a_8^2$					
$x_3 x_5 x_9 a_4 a_{10}$	0	0	0	$x_3 x_5 x_9 a_6 a_{10}$		
$x_3 x_5 x_9 a_6 a_8$	$x_3^3 x_5 x_9 a_6 a_8$					
$x_3 x_7 x_9 a_4 a_8$	$x_3^3 x_7 x_9 a_4 a_8$					
$x_3 x_7 x_9 a_6^2$	$x_3^3 x_7 x_9 a_6^2$					
$x_5 x_7 x_9 a_4 a_6$	$x_3^2 x_5 x_7 x_9 a_4 a_6$					

$\mathrm{Cotor}^{3,28}$	x_3^2	x_9	Sq^1	Sq^2	Sq^4	Sq^8
$x_3^3 x_5 a_4 a_6 a_7$	0	0	$x_3^3 x_5 a_4 a_7^2$			
$x_3^3 x_7 a_4^2 a_7$	0	0	0	$x_3^2 x_5 x_7 a_4^2 a_7$		
$x_3^3 a_4 a_8 a_{10}$	0	0	0	$x_3^2 x_5 a_4 a_8 a_{10} + x_3^3 a_6 a_8 a_{10}$		
$x_3^3 a_6^2 a_{10}$	0	0	0	$x_3^3 x_5 a_6^2 a_{10}$	$x_3^3 a_8^2 a_{10}$	
$x_3^3 a_6 a_8^2$	0	$x_3^3 x_9 a_6 a_8^2$				
$x_3^3 a_7^2 a_8$	0	0	0	$x_3^2 x_5 a_7^2 a_8$		
$x_3^2 x_5 a_4 a_6 a_{10}$	0	0	0	$x_3^2 x_5 a_6^2 a_{10}$	0	
$x_3^2 x_5 a_4 a_8^2$	0	$x_3^2 x_5 x_9 a_4 a_8^2$				
$x_3^2 x_5 a_6^2 a_8$	0	$x_3^2 x_5 x_9 a_6^2 a_8$				
$x_3^2 x_5 a_6 a_7^2$	0	0	$x_3^2 x_5 a_7^3$			
$x_3^2 x_7 a_4^2 a_{10}$	0	0	0	0	$x_3^2 x_7 a_6^2 a_{10}$	
$x_3^2 x_7 a_4 a_6 a_8$	0	$x_3^2 x_7 x_9 a_4 a_6 a_8$				

13. APPENDIX

Cotor^{3,28}	x_3^2	x_9	Sq^1	Sq^2	Sq^4	Sq^8
$x_3^2 x_7 a_4 a_7^2$	0	0	0	$x_3^2 x_7 a_6 a_7^2$		
$x_3^2 x_7 a_6^3$	0	$x_3^2 x_7 x_9 a_6^3$				
$x_3^2 x_9 a_4^2 a_8$	0	0	0	0	$x_3^2 x_9 a_6^2 a_8$	
$x_3^2 x_9 a_4 a_6^2$	0	0	0	$x_3^2 x_9 a_6^3$		
$x_3 x_5 x_7 a_4^2 a_8$	$x_3^3 x_5 x_7 a_4^2 a_8$					
$x_3 x_5 x_7 a_4 a_6^2$	$x_3^3 x_5 x_7 a_4 a_6^2$					
$x_3 x_5 x_9 a_4^2 a_6$	$x_3^3 x_5 x_9 a_4^2 a_6$					
$x_3 x_5 a_7 a_8^2$	$x_3^3 x_5 a_7 a_8^2$					
$x_3 x_7 x_9 a_4^3$	$x_3^3 x_7 x_9 a_4^3$					
$x_3 x_7 a_6 a_7 a_8$	$x_3^3 x_7 a_6 a_7 a_8$					
$x_3 x_7 a_7^3$	$x_3^3 x_7 a_7^3$					
$x_3 a_8 a_{10}^2$	0	$x_3 x_9 a_8 a_{10}^2$				
$x_5 x_7 a_4 a_7 a_8$	$x_3^2 x_5 x_7 a_4 a_7 a_8$					
$x_5 x_7 a_6^2 a_7$	$x_3^2 x_5 x_7 a_6^2 a_7$					
$x_5 a_6 a_{10}^2$	0	$x_5 x_9 a_6 a_{10}^2$				
$x_5 a_8^2 a_{10}$	$x_3^2 x_5 a_8^2 a_{10}$					
$x_7 a_4 a_{10}^2$	0	$x_7 x_9 a_4 a_{10}^2$				
$x_7 a_6 a_8 a_{10}$	$x_3^2 x_7 a_6 a_8 a_{10}$					
$x_7 a_8^3$	$x_3^2 x_7 a_8^3$					
$x_9 a_4 a_8 a_{10}$	0	0	0	$x_9 a_6 a_8 a_{10} + x_9 a_4 a_{10}^2$		
$x_9 a_6^2 a_{10}$	0	0	0	0	0	$x_9 a_{10}^3$
$x_9 a_6 a_8^2$	$x_3^2 x_9 a_6 a_8^2$					

Cotor^{4,27}	x_3^2	x_9	Sq^1	Sq^2	Sq^4	Sq^8
$x_3^3 a_4^3 a_{10}$	0	0	0	$x_3^2 x_5 a_4^3 a_{10} + x_3^3 a_4^2 a_6 a_{10}$		
$x_3^3 a_4^2 a_6 a_8$	0	$x_3^3 x_9 a_4^2 a_6 a_8$				
$x_3^3 a_4^2 a_7^2$	0	0	0	$x_3^2 x_5 a_4^2 a_7^2$		
$x_3^3 a_4 a_6^3$	0	$x_3^3 x_9 a_4 a_6^3$				
$x_3^2 x_5 a_4^3 a_8$	0	$x_3^2 x_5 x_9 a_4^3 a_8$				
$x_3^2 x_5 a_4^2 a_6^2$	0	$x_3^2 x_5 x_9 a_4^2 a_6^2$				
$x_3^2 x_7 a_4^3 a_6$	0	$x_3^2 x_7 x_9 a_4^3 a_6$				
$x_3^2 x_9 a_4^4$	0	0	0	0	0	$x_3^2 x_9 a_6^4$
$x_3^2 a_4 a_6 a_7 a_8$	0	0	$x_3^2 a_4 a_7^2 a_8$			
$x_3^2 a_4 a_7^3$	0	0	0	$x_3^2 a_6 a_7^3$		
$x_3^2 a_6^3 a_7$	0	0	$x_3^2 a_6^2 a_7^2$			
$x_3 x_5 x_7 a_4^4$	$x_3^3 x_5 x_7 a_4^4$					
$x_3 x_5 a_4^2 a_7 a_8$	$x_3^3 x_5 a_4^2 a_7 a_8$					

$\text{Cotor}^{4,27}$	x_3^2	x_9	Sq^1	Sq^2	Sq^4	Sq^8
$x_3 x_5 a_4 a_6^2 a_7$	$x_3^3 x_5 a_4 a_6^2 a_7$					
$x_3 x_7 a_4^2 a_6 a_7$	$x_3^3 x_7 a_4^2 a_6 a_7$					
$x_3 a_4^2 a_{10}^2$	0	$x_3 x_9 a_4^2 a_{10}^2$				
$x_3 a_4 a_6 a_8 a_{10}$	$x_3^3 a_4 a_6 a_8 a_{10}$					
$x_3 a_4 a_8^3$	$x_3^3 a_4 a_8^3$					
$x_3 a_6^3 a_{10}$	$x_3^3 a_6^3 a_{10}$					
$x_3 a_6^2 a_8^2$	$x_3^3 a_6^2 a_8^2$					
$x_3 a_6 a_7^2 a_8$	$x_3^3 a_6 a_7^2 a_8$					
$x_3 a_7^4$	$x_3^3 a_7^4$					
$x_5 x_7 a_4^3 a_7$	$x_3^2 x_5 x_7 a_4^3 a_7$					
$x_5 a_4^2 a_8 a_{10}$	$x_3^2 x_5 a_4^2 a_8 a_{10}$					
$x_5 a_4 a_6^2 a_{10}$	$x_3^2 x_5 a_4 a_6^2 a_{10}$					
$x_5 a_4 a_6 a_8^2$	$x_3^2 x_5 a_4 a_6 a_8^2$					
$x_5 a_4 a_7^2 a_8$	$x_3^2 x_5 a_4 a_7^2 a_8$					
$x_5 a_6^3 a_8$	$x_3^2 x_5 a_6^3 a_8$					
$x_5 a_6^2 a_7^2$	$x_3^2 x_5 a_6^2 a_7^2$					
$x_7 a_4^2 a_6 a_{10}$	$x_3^2 x_7 a_4^2 a_6 a_{10}$					
$x_7 a_4^2 a_8^2$	$x_3^2 x_7 a_4^2 a_8^2$					
$x_7 a_4 a_6^2 a_8$	$x_3^2 x_7 a_4 a_6^2 a_8$					
$x_7 a_4 a_6 a_7^2$	$x_3^2 x_7 a_4 a_6 a_7^2$					
$x_7 a_6^4$	$x_3^2 x_7 a_6^4$					
$x_9 a_4^3 a_{10}$	0	0	0	$x_9 a_4^2 a_6 a_{10}$		
$x_9 a_4^2 a_6 a_8$	$x_3^2 x_9 a_4^2 a_6 a_8$					
$x_9 a_4 a_6^3$	$x_3^2 x_9 a_4 a_6^3$					
$a_7 a_8^3$	$x_3^2 a_7 a_8^3$					

$\text{Cotor}^{5,26}$	x_3^2	x_9	Sq^1	Sq^2	Sq^4	Sq^8
$x_3^3 a_4^4 a_6$	0	$x_3^3 x_9 a_4^4 a_6$				
$x_3^2 x_5 a_4^5$	0	$x_3^2 x_5 x_9 a_4^5$				
$x_3^2 a_4^3 a_6 a_7$	0	0	$x_3^2 a_4^3 a_7^2$			
$x_3 x_5 a_4^4 a_7$	$x_3^3 x_5 a_4^4 a_7$					
$x_3 a_4^3 a_6 a_{10}$	$x_3^3 a_4^3 a_6 a_{10}$					
$x_3 a_4^3 a_8^2$	$x_3^3 a_4^3 a_8^2$					
$x_3 a_4^2 a_6^2 a_8$	$x_3^3 a_4^2 a_6^2 a_8$					
$x_3 a_4^2 a_6 a_7^2$	$x_3^3 a_4^2 a_6 a_7^2$					
$x_3 a_4 a_6^4$	$x_3^3 a_4 a_6^4$					
$x_5 a_4^4 a_{10}$	$x_3^2 x_5 a_4^4 a_{10}$					

13. APPENDIX

Cotor^{5,26}	x_3^2	x_9	Sq^1	Sq^2	Sq^4	Sq^8
$x_5 a_4^3 a_6 a_8$	$x_3^2 x_5 a_4^3 a_6 a_8$					
$x_5 a_4^3 a_7^2$	$x_3^2 x_5 a_4^3 a_7^2$					
$x_5 a_4^2 a_6^3$	$x_3^2 x_5 a_4^2 a_6^3$					
$x_7 a_4^4 a_8$	$x_3^2 x_7 a_4^4 a_8$					
$x_7 a_4^3 a_6^2$	$x_3^2 x_7 a_4^3 a_6^2$					
$x_9 a_4^4 a_6$	$x_3^2 x_9 a_4^4 a_6$					
$a_4^2 a_7 a_8^2$	$x_3^2 a_4^2 a_7 a_8^2$					
$a_4 a_6^2 a_7 a_8$	$x_3^2 a_4 a_6^2 a_7 a_8$					
$a_4 a_6 a_7^3$	$x_3^2 a_4 a_6 a_7^3$					
$a_6^4 a_7$	$x_3^2 a_6^4 a_7$					

Cotor^{6,25}	x_3^2	x_9	Sq^1	Sq^2	Sq^4	Sq^8
$x_3 a_4^5 a_8$	$x_3^3 a_4^5 a_8$					
$x_3 a_4^4 a_6^2$	$x_3^3 a_4^4 a_6^2$					
$x_5 a_4^5 a_6$	$x_3^2 x_5 a_4^5 a_6$					
$x_7 a_4^6$	$x_3^2 x_7 a_4^6$					
$a_4^4 a_7 a_8$	$x_3^2 a_4^4 a_7 a_8$					
$a_4^3 a_6^2 a_7$	$x_3^2 a_4^3 a_6^2 a_7$					

Cotor^{7,24}	x_3^2	x_9	Sq^1	Sq^2	Sq^4	Sq^8
$x_3 a_4^7$	$x_3^3 a_4^7$					
$a_4^6 a_7$	$x_3^2 a_4^6 a_7$					

Bibliography

[1] J. F. Adams, Lectures on Exceptional Lie Groups, (with a foreword by J. Peter May, edited by Zafer Mahmud and Mamoru Mimura) Chicago Lectures in Mathematics. University of Chicago Press, Chicago, IL, 1996.

[2] A. Borel, Sur l'homologie et la cohomologie des groupes de Lie compacts connexes, Amer. J. Math. **76**(1954), 273–342.

[3] E. H. Brown, Jr., Twisted tensor products, I, Ann. Math., **69**(1959), 223-246.

[4] N. Castellana and N. Kitchloo, A homotopy construction of the adjoint representation for Lie groups. Math. Proc. Cambridge Philos. Soc. **133**(2002), 399–409.

[5] M. C. Crabb and W.A.Sutherland, Counting homotopy types of gauge groups, Proc. London Math. Soc. (3), **81**(2000), 747-768.

[6] S. Eilenberg and J. C. Moore, Homology and fibrations, coalgebras, cotensor product, and its derived functors, Comm. Math. Helv., **40**(1966), 199-236.

[7] S. I. Gelfand and Yu. I. Manin, Methods of Homological Algebras, Berlin-Heidelberg: Springer-Verlag, 1996.

[8] E. Getzler and J. D. S. Jones, A_∞- algebras and cyclic bar complex, Illinois J. Math., **34**(1990), 256-283.

[9] E. Getzler, J. D. S. Jones and S. Petrack, Differential form on loop spaces and the cyclic bar complex, Topology, **30**(1991), 339-371.

[10] S. Halperin and J. M. Lemaire, Notions of category in differential algebra, Algebraic Topology: Rational Homotopy, Springer Lecture Notes in Math., **1318**, Springer, Berlin, New York, 1988, pp. 138-154.

[11] H. Hamanaka, Homology ring mod 2 of free loop groups of exceptional Lie groups, J. Math. Kyoto Univ., **36**(1996), 669-686.

[12] H. Hamanaka, Homology ring mod 2 of free loop groups of spinor groups, J. Pure Appl. Algebra, **146**(2000), 267-282.

[13] H. Hamanaka and S. Hara, The mod 3 homology of the space of loops on the exceptional Lie groups and the adjoint action, J. Math. Kyoto Univ., **37**(1997), 441-453.

[14] H. Hamanaka, S. Hara and A. Kono, Adjoint actions on the modulo 5 homology groups of E_8 and ΩE_8, J. Math. Kyoto Univ., **37**(1997), 169-176.

[15] K. Hess, Twisted tensor products of DGA's and Adams-Hilton model for the total space of a fibration, London Math. Soc. Lecture Note Series, **175**(1992), 29–51.

[16] J. Hunton, M. Mimura, T. Nishimoto and B. Schuster, Higher v_n-torsion in Lie groups, J. Math. Soc. Japan, **50**(1998), 801–818.

[17] K. Ishitoya, A. Kono and H. Toda, Hopf algebra structure of mod 2 cohomology of simple Lie groups, Publ. RIMS Kyoto Univ., **12**(1976/77), 141–167.

[18] J. D. S. Jones, Cyclic homology and equivariant homology, Invent. Math., **87**(1987), 403-423.

[19] S. Kleinerman, The cohomology of Chevalley groups of exceptional Lie type, Memoirs of AMS, **268**(1982).

[20] K. Kono and K. Kozima, The adjoint action of a Lie group on the space on loops, J. Math. Soc. Japan, **45**(1993), 495-510.

[21] A. Kono and K. Kuribayashi, Module derivations and cohomological splitting of adjoint bundles, submitted to Fundamenta Math., (2002).

[22] A. Kono and M. Mimura, Cohomology mod 2 of the classifying space of the compact connected Lie group of type E_6, J. Pure and Applied Algebra, **6**(1975), 61-81.

[23] A. Kono, M. Mimura and N. Shimada, Cohomology of classifying space of certain associative H-spaces, J. Math. Kyoto Univ., **15**(1975), 607-617.

[24] A. Kono and M. Mimura and N. Shimada, On the cohomology mod 2 of the classifying space of the 1-connected exceptional Lie group E_7, J. Pure and Applied Algebra, **8**(1976), 267-283.

[25] D. Kraines and C. Schochet, Differentials in the Eilenberg-Moore spectral sequence, J. Pure and Applied Algebra, **2**(1972), 131-148.

[26] K. Kuribayashi, On the mod p cohomology of spaces of free loops on the Grassmann and Stiefel manifolds, J. Math. Soc. Japan, **43**(1991), 331-346.

[27] K. Kuribayashi, On the real cohomology of spaces of free loops on manifolds, Fundamenta Math., **150**(1996), 173-188.

[28] K. Kuribayashi, Module derivations and the adjoint action of a finite loop space, J. Math. Kyoto Univ., **39**(1999), 67-85.

[29] R. J. Milgram and M. Tezuka, The geometry and cohomology of M_{12} :II, Bol. Soc. Mat. Mexicana(3), **1**(1995), 91-108.

[30] J. W. Milnor and J. C. Moore, On the structure of Hopf algebras, Ann. of Math., **81**(1965), 211-236.

[31] M. Mimura, The characteristic classes for the exceptional Lie groups, London Math. Soc. Lecture Note Series, **175**(1992), 103-130.

[32] M. Mimura and Y. Sambe, On the cohomology mod p of the classifying spaces of the exceptional Lie groups, I, J. Math. Kyoto Univ., **19**(1979), 553-581.

[33] M. Mimura and Y. Sambe, On the cohomology mod p of the classifying spaces of the exceptional Lie groups, II, J. Math. Kyoto Univ., **20**(1980), 327-349.

[34] M. Mimura and Y. Sambe, Collapsing of the Eilenberg-Moore spectral sequence mod 5 of the compact exceptional group E_8, J. Math. Kyoto Univ., **21**(1981), 203-230.

[35] M. Mori, The Steenrod operations in the Eilenberg-Moore spectral sequence, Hiroshima Math. J., **9**(1979), 17-34.

[36] H. J. Munkholm, The Eilenberg-Moore spectral sequence and strongly homotopy multiplicative maps, J. Pure and Applied Algebra, **5**(1974), 1-50.

[37] B. Ndombol and J. -C. Thomas, On the cohomology algebra of free loop spaces, Topology, **41**(2002), 85-106.

[38] T. Nishimoto, Higher torsion in the Morava K-theory of $SO(m)$ and $Spin(m)$, J. Math. Soc. Japan, **53**(2001), 383-394.

[39] D. Quillen, The mod 2 cohomology ring of extra-special 2-groups and spinor groups, Math. Ann., **194**(1971), 197-212.

[40] D. C. Ravenel, Complex Cobordism and Stable Homotopy Groups of Spheres, Academic Press, 1986.

[41] N. Shimada and A. Iwai, On the cohomology of some Hopf algebras, Nagoya Math. J., **30**(1967), 103-111.

[42] L. Smith, On the characteristic zero cohomology of free loop space, Amer. J. Math., **103**(1981), 887-910.

[43] L. Smith, The Eilenberg-Moore spectral sequence and the mod 2 cohomology of certain free loop spaces, Illinois J. Math., **28**(1984), 516-522.

[44] A. Vavpetič, Homotopy characterization of classifying spaces of compact Lie groups, (Thesis Univ. of Ljubljana, 2000)

[45] C. Weibel, An Introduction to Homological Algebra. Cambridge Studies in Advanced Mathematics **38** Cambridge: Cambridge University Press, 1994.

[46] G. W. Whitehead, Elements of Homotopy Theory, Graduate Texts in Mathematics, 61. Springer-Verlag, New York-Berlin, 1978.

[47] I. Yokota, Exceptional Lie group F_4 and its representation rings, J. Fac. Sci. Shinshu Univ., **3**(1968), 35–60.

Editorial Information

To be published in the *Memoirs*, a paper must be correct, new, nontrivial, and significant. Further, it must be well written and of interest to a substantial number of mathematicians. Piecemeal results, such as an inconclusive step toward an unproved major theorem or a minor variation on a known result, are in general not acceptable for publication. Papers appearing in *Memoirs* are generally at least 80 and not more than 200 published pages in length. Papers less than 80 or more than 200 published pages require the approval of the Managing Editor of the Transactions/Memoirs Editorial Board.

As of November 30, 2005, the backlog for this journal was approximately 15 volumes. This estimate is the result of dividing the number of manuscripts for this journal in the Providence office that have not yet gone to the printer on the above date by the average number of monographs per volume over the previous twelve months, reduced by the number of volumes published in four months (the time necessary for preparing a volume for the printer). (There are 6 volumes per year, each containing at least 4 numbers.)

A Consent to Publish and Copyright Agreement is required before a paper will be published in the *Memoirs*. After a paper is accepted for publication, the Providence office will send a Consent to Publish and Copyright Agreement to all authors of the paper. By submitting a paper to the *Memoirs*, authors certify that the results have not been submitted to nor are they under consideration for publication by another journal, conference proceedings, or similar publication.

Information for Authors

Memoirs are printed from camera copy fully prepared by the author. This means that the finished book will look exactly like the copy submitted.

The paper must contain a *descriptive title* and an *abstract* that summarizes the article in language suitable for workers in the general field (algebra, analysis, etc.). The *descriptive title* should be short, but informative; useless or vague phrases such as "some remarks about" or "concerning" should be avoided. The *abstract* should be at least one complete sentence, and at most 300 words. Included with the footnotes to the paper should be the 2000 *Mathematics Subject Classification* representing the primary and secondary subjects of the article. The classifications are accessible from www.ams.org/msc/. The list of classifications is also available in print starting with the 1999 annual index of *Mathematical Reviews*. The Mathematics Subject Classification footnote may be followed by a list of *key words and phrases* describing the subject matter of the article and taken from it. Journal abbreviations used in bibliographies are listed in the latest *Mathematical Reviews* annual index. The series abbreviations are also accessible from www.ams.org/publications/. To help in preparing and verifying references, the AMS offers MR Lookup, a Reference Tool for Linking, at www.ams.org/mrlookup/. When the manuscript is submitted, authors should supply the editor with electronic addresses if available. These will be printed after the postal address at the end of the article.

Electronically prepared manuscripts. The AMS encourages electronically prepared manuscripts, with a strong preference for $\mathcal{AMS}$-LaTeX. To this end, the Society has prepared $\mathcal{AMS}$-LaTeX author packages for each AMS publication. Author packages include instructions for preparing electronic manuscripts, the *AMS Author Handbook*, samples, and a style file that generates the particular design specifications of that publication series. Though $\mathcal{AMS}$-LaTeX is the highly preferred format of TeX, author packages are also available in $\mathcal{AMS}$-TeX.

Authors may retrieve an author package from e-MATH starting from www.ams.org/tex/ or via FTP to ftp.ams.org (login as anonymous, enter username as password, and type cd pub/author-info). The *AMS Author Handbook* and the *Instruction Manual* are available in PDF format following the author packages link from www.ams.org/tex/. The author package can be obtained free of charge by sending email to pub@ams.org (Internet) or from the Publication Division, American Mathematical Society, 201 Charles St., Providence, RI 02904, USA. When requesting an author package, please specify $\mathcal{AMS}$-LaTeX or $\mathcal{AMS}$-TeX, Macintosh or IBM (3.5) format, and the publication in which your paper will appear. Please be sure to include your complete mailing address.

Sending electronic files. After acceptance, the source file(s) should be sent to the Providence office (this includes any TeX source file, any graphics files, and the DVI or PostScript file).

Before sending the source file, be sure you have proofread your paper carefully. The files you send must be the EXACT files used to generate the proof copy that was accepted for publication. For all publications, authors are required to send a printed copy of their paper, which exactly matches the copy approved for publication, along with any graphics that will appear in the paper.

TeX files may be submitted by email, FTP, or on diskette. The DVI file(s) and PostScript files should be submitted only by FTP or on diskette unless they are encoded properly to submit through email. (DVI files are binary and PostScript files tend to be very large.)

Electronically prepared manuscripts can be sent via email to pub-submit@ams.org (Internet). The subject line of the message should include the publication code to identify it as a Memoir. TeX source files, DVI files, and PostScript files can be transferred over the Internet by FTP to the Internet node e-math.ams.org (130.44.1.100).

Electronic graphics. Comprehensive instructions on preparing graphics are available at www.ams.org/jourhtml/graphics.html. A few of the major requirements are given here.

Submit files for graphics as EPS (Encapsulated PostScript) files. This includes graphics originated via a graphics application as well as scanned photographs or other computer-generated images. If this is not possible, TIFF files are acceptable as long as they can be opened in Adobe Photoshop or Illustrator. No matter what method was used to produce the graphic, it is necessary to provide a paper copy to the AMS.

Authors using graphics packages for the creation of electronic art should also avoid the use of any lines thinner than 0.5 points in width. Many graphics packages allow the user to specify a "hairline" for a very thin line. Hairlines often look acceptable when proofed on a typical laser printer. However, when produced on a high-resolution laser imagesetter, hairlines become nearly invisible and will be lost entirely in the final printing process.

Screens should be set to values between 15% and 85%. Screens which fall outside of this range are too light or too dark to print correctly. Variations of screens within a graphic should be no less than 10%.

Inquiries. Any inquiries concerning a paper that has been accepted for publication should be sent directly to the Electronic Prepress Department, American Mathematical Society, 201 Charles St., Providence, RI 02904, USA.

Editors

This journal is designed particularly for long research papers, normally at least 80 pages in length, and groups of cognate papers in pure and applied mathematics. Papers intended for publication in the *Memoirs* should be addressed to one of the following editors. In principle the Memoirs welcomes electronic submissions, and some of the editors, those whose names appear below with an asterisk (*), have indicated that they prefer them. However, editors reserve the right to request hard copies after papers have been submitted electronically. Authors are advised to make preliminary email inquiries to editors about whether they are likely to be able to handle submissions in a particular electronic form.

*Algebra to ALEXANDER KLESHCHEV, Department of Mathematics, University of Oregon, Eugene, OR 97403-1222; email: ams@noether.uoregon.edu

Algebra and its application to MINA TEICHER, Emmy Noether Research Institute for Mathematics, Bar-Ilan University, Ramat-Gan 52900, Israel; email: teicher@macs.biu.ac.il

Algebraic geometry to DAN ABRAMOVICH, Department of Mathematics, Brown University, Box 1917, Providence, RI 02912; email: amsedit@math.brown.edu

*Algebraic number theory to V. KUMAR MURTY, Department of Mathematics, University of Toronto, 100 St. George Street, Toronto, ON M5S 1A1, Canada; email: murty@math.toronto.edu

*Algebraic topology to ALEJANDRO ADEM, Department of Mathematics, University of British Columbia, Room 121, 1984 Mathematics Road, Vancouver, British Columbia, Canada V6T 1Z2; email: adem@math.ubc.ca

Combinatorics to JOHN R. STEMBRIDGE, Department of Mathematics, University of Michigan, Ann Arbor, Michigan 48109-1109; email: jrs@umich.edu

Complex analysis and harmonic analysis to ALEXANDER NAGEL, Department of Mathematics, University of Wisconsin, 480 Lincoln Drive, Madison, WI 53706-1313; email: nagel@math.wisc.edu

*Differential geometry and global analysis to LISA C. JEFFREY, Department of Mathematics, University of Toronto, 100 St. George St., Toronto, ON Canada M5S 3G3; email: jeffrey@math.toronto.edu

Dynamical systems and ergodic theory to AMIE WILKINSON, Department of Mathematics, Northwestern University, 2033 Sheridan Road, Evanston, IL 60208-2730; email: wilkinso@math.northwestern.edu

*Functional analysis and operator algebras to MARIUS DADARLAT, Department of Mathematics, Purdue University, 150 N. University St., West Lafayette, IN 47907-2067; email: mdd@math.purdue.edu

*Geometric analysis to TOBIAS COLDING, Courant Institute, New York University, 251 Mercer St., New York, NY 10012; email: traneditor@cims.nyu.edu

*Geometric analysis to MLADEN BESTVINA, Department of Mathematics, University of Utah, 155 South 1400 East, JWB 233, Salt Lake City, Utah 84112-0090; email: bestvina@math.utah.edu

Harmonic analysis, representation theory, and Lie theory to ROBERT J. STANTON, Department of Mathematics, The Ohio State University, 231 West 18th Avenue, Columbus, OH 43210-1174; email: stanton@math.ohio-state.edu

*Logic to STEFFEN LEMPP, Department of Mathematics, University of Wisconsin, 480 Lincoln Drive, Madison, Wisconsin 53706-1388; email: lempp@math.wisc.edu

*Ordinary differential equations, and applied mathematics to PETER W. BATES, Department of Mathematics, Michigan State University, East Lansing, MI 48824-1027; email: bates@math.msu.edu

*Partial differential equations to GUSTAVO PONCE, Department of Mathematics, South Hall, Room 6607, University of California, Santa Barbara, CA 93106; email: ponce@math.ucsb.edu

*Probability and statistics to KRZYSZTOF BURDZY, Department of Mathematics, University of Washington, Box 354350, Seattle, Washington 98195-4350; email: burdzy@math.washington.edu

*Real analysis and partial differential equations to DANIEL TATARU, Department of Mathematics, University of California, Berkeley, Berkeley, CA 94720; email: tataru@math.berkeley.edu

All other communications to the editors should be addressed to the Managing Editor, ROBERT GURALNICK, Department of Mathematics, University of Southern California, Los Angeles, CA 90089-1113; email: guralnic@math.usc.edu.

Titles in This Series

851 **Jie Wu,** On maps from loop suspensions to loop spaces and the shuffle relations on the Cohen groups, 2006

850 **Siegfried Echterhoff, S. Kaliszewski, John Quigg, and Iain Raeburn,** A categorical approach to imprimitivity theorems for C^*-dynamical systems, 2006

849 **Katsuhiko Kuribayashi, Mamoru Mimura, and Tetsu Nishimoto,** Twisted tensor products related to the cohomology of the classifying spaces of loop groups, 2006

848 **Bob Oliver,** Equivalences of classifying spaces completed at the prime two, 2006

847 **Eric T. Sawyer and Richard L. Wheeden,** Hölder continuity of weak solutions to subelliptic equations with rough coefficients, 2006

846 **Victor Beresnevich, Detta Dickinson, and Sanju Velani,** Measure theoretic laws for lim–sup sets, 2006

845 **Ehud Friedgut, Vojtech Rödl, Andrzej Ruciński, and Prasad V. Tetali,** A Sharp threshold for random graphs with a monochromatic triangle in every edge coloring, 2006

844 **Amadeu Delshams, Rafael de la Llave, and Tere M. Seara,** A geometric mechanism for diffusion in Hamiltonian systems overcoming the large gap problem: Heuristics and rigorous verification on a model, 2006

843 **Denis V. Osin,** Relatively hyperbolic groups: Intrinsic geometry, algebraic properties, and algorithmic problems, 2006

842 **David P. Blecher and Vrej Zarikian,** The calculus of one-sided M-ideals and multipliers in operator spaces, 2006

841 **Enrique Artal Bartolo, Pierrette Cassou-Noguès, Ignacio Luengo, and Alejandro Melle Hernández,** Quasi-ordinary power series and their zeta functions, 2005

840 **Sławomir Kołodziej,** The complex Monge-Ampère equation and pluripotential theory, 2005

839 **Mihai Ciucu,** A random tiling model for two dimensional electrostatics, 2005

838 **V. Jurdjevic,** Integrable Hamiltonian systems on complex Lie groups, 2005

837 **Joseph A. Ball and Victor Vinnikov,** Lax-Phillips scattering and conservative linear systems: A Cuntz-algebra multidimensional setting, 2005

836 **H. G. Dales and A. T.-M. Lau,** The second duals of Beurling algebras, 2005

835 **Kiyoshi Igusa,** Higher complex torsion and the framing principle, 2005

834 **Kenichi Ohshika,** Kleinian groups which are limits of geometrically finite groups, 2005

833 **Greg Hjorth and Alexander S. Kechris,** Rigidity theorems for actions of product groups and countable Borel equivalence relations, 2005

832 **Lee Klingler and Lawrence S. Levy,** Representation type of commutative Noetherian rings III: Global wildness and tameness, 2005

831 **K. R. Goodearl and F. Wehrung,** The complete dimension theory of partially ordered systems with equivalence and orthogonality, 2005

830 **Jason Fulman, Peter M. Neumann, and Cheryl E. Praeger,** A generating function approach to the enumeration of matrices in classical groups over finite fields, 2005

829 **S. G. Bobkov and B. Zegarlinski,** Entropy bounds and isoperimetry, 2005

828 **Joel Berman and Paweł M. Idziak,** Generative complexity in algebra, 2005

827 **Trevor A. Welsh,** Fermionic expressions for minimal model Virasoro characters, 2005

826 **Guy Métivier and Kevin Zumbrun,** Large viscous boundary layers for noncharacteristic nonlinear hyperbolic problems, 2005

825 **Yaozhong Hu,** Integral transformations and anticipative calculus for fractional Brownian motions, 2005

824 **Luen-Chau Li and Serge Parmentier,** On dynamical Poisson groupoids I, 2005

TITLES IN THIS SERIES

823 **Claus Mokler,** An analogue of a reductive algebraic monoid whose unit group is a Kac-Moody group, 2005

822 **Stefano Pigola, Marco Rigoli, and Alberto G. Setti,** Maximum principles on Riemannian manifolds and applications, 2005

821 **Nicole Bopp and Hubert Rubenthaler,** Local zeta functions attached to the minimal spherical series for a class of symmetric spaces, 2005

820 **Vadim A. Kaimanovich and Mikhail Lyubich,** Conformal and harmonic measures on laminations associated with rational maps, 2005

819 **F. Andreatta and E. Z. Goren,** Hilbert modular forms: Mod p and p-adic aspects, 2005

818 **Tom De Medts,** An algebraic structure for Moufang quadrangles, 2005

817 **Javier Fernández de Bobadilla,** Moduli spaces of polynomials in two variables, 2005

816 **Francis Clarke,** Necessary conditions in dynamic optimization, 2005

815 **Martin Bendersky and Donald M. Davis,** V_1-periodic homotopy groups of $SO(n)$, 2004

814 **Johannes Huebschmann,** Kähler spaces, nilpotent orbits, and singular reduction, 2004

813 **Jeff Groah and Blake Temple,** Shock-wave solutions of the Einstein equations with perfect fluid sources: Existence and consistency by a locally inertial Glimm scheme, 2004

812 **Richard D. Canary and Darryl McCullough,** Homotopy equivalences of 3-manifolds and deformation theory of Kleinian groups, 2004

811 **Ottmar Loos and Erhard Neher,** Locally finite root systems, 2004

810 **W. N. Everitt and L. Markus,** Infinite dimensional complex symplectic spaces, 2004

809 **J. T. Cox, D. A. Dawson, and A. Greven,** Mutually catalytic super branching random walks: Large finite systems and renormalization analysis, 2004

808 **Hagen Meltzer,** Exceptional vector bundles, tilting sheaves and tilting complexes for weighted projective lines, 2004

807 **Carlos A. Cabrelli, Christopher Heil, and Ursula M. Molter,** Self-similarity and multiwavelets in higher dimensions, 2004

806 **Spiros A. Argyros and Andreas Tolias,** Methods in the theory of hereditarily indecomposable Banach spaces, 2004

805 **Philip L. Bowers and Kenneth Stephenson,** Uniformizing dessins and Belyĭ maps via circle packing, 2004

804 **A. Yu Ol'shanskii and M. V. Sapir,** The conjugacy problem and Higman embeddings, 2004

803 **Michael Field and Matthew Nicol,** Ergodic theory of equivariant diffeomorphisms: Markov partitions and stable ergodicity, 2004

802 **Martin W. Liebeck and Gary M. Seitz,** The maximal subgroups of positive dimension in exceptional algebraic groups, 2004

801 **Fabio Ancona and Andrea Marson,** Well-posedness for general 2×2 systems of conservation law, 2004

800 **V. Poénaru and C. Tanas,** Equivariant, almost-arborescent representation of open simply-connected 3-manifolds; A finiteness result, 2004

799 **Barry Mazur and Karl Rubin,** Kolyvagin systems, 2004

798 **Benoît Mselati,** Classification and probabilistic representation of the positive solutions of a semilinear elliptic equation, 2004

797 **Ola Bratteli, Palle E. T. Jorgensen, and Vasyl' Ostrovs'kyĭ,** Representation theory and numerical AF-invariants, 2004

For a complete list of titles in this series, visit the
AMS Bookstore at **www.ams.org/bookstore/**.